Lecture Notes
in Business Information Processing

429

More information about this series at http://www.springer.com/series/7911

Stanisław Wrycza · Jacek Maślankowski (Eds.)

Digital Transformation

13th PLAIS EuroSymposium
on Digital Transformation, PLAIS EuroSymposium 2021
Sopot, Poland, September 23, 2021
Proceedings

 Springer

Editors
Stanisław Wrycza ⓘ
University of Gdansk
Sopot, Poland

Jacek Maślankowski ⓘ
University of Gdansk
Sopot, Poland

ISSN 1865-1348   ISSN 1865-1356   (electronic)
Lecture Notes in Business Information Processing
ISBN 978-3-030-85892-6   ISBN 978-3-030-85893-3   (eBook)
https://doi.org/10.1007/978-3-030-85893-3

This Springer imprint is published by the registered company Springer Nature Switzerland AG
The registered company address is: Gewerbestrasse 11, 6330 Cham, Switzerland

Preface

The annual academic event EuroSymposium—which has reached the thirteenth edition—has been changing and extending its thematic scope, according to the quick progress of information systems (IS) knowledge and applications, in the era of the digital transformation. The new emerging areas are included in the list of the event's topics: digital science, digital enterprises, smart cities, innovative methods of data and process analysis, user experience (UX) design, digital education, deep learning, social media use and analytics, the Internet of Things (IoT), and big data. All of these areas have had a strong influence on the EuroSymposium subject matter, changing it fundamentally, especially when compared with the early EuroSymposia.

The objective of the EuroSymposium is to promote and develop high quality research on all issues related to digital transformation. It provides a forum for IS researchers and practitioners in Europe and beyond to interact, collaborate, and develop this field. The EuroSymposia were initiated by Keng Siau as part of the SIGSAND-Europe Initiative. Previous EuroSymposia were held at the University of Galway, Ireland (2006), the University of Gdańsk, Poland (2007), the University of Marburg, Germany (2008), and the University of Gdańsk, Poland (2011–2019). The papers accepted for presentation at the EuroSymposia were published in the following proceedings:

- 2nd EuroSymposium 2007: Bajaj, A, Wrycza, S (eds) (2009) Systems Analysis and Design for Advanced Modeling Methods: Best Practices. IGI Global, Hershey, PA
- 4th EuroSymposium 2011: Wrycza, S (ed) (2011) Research in Systems Analysis and Design: Models and Methods, LNBIP 93. Springer-Verlag, Berlin
- Joint Working Conferences EMMSAD/5th EuroSymposium held at CAiSE 2012: Bider, I, Halpin, T, Krogstie, J, Nurcan, S, Proper, E, Schmidt, R, Soffer, P, Wrycza, S (eds) (2012) Enterprise, Business-Process and Information Systems Modeling, LNBIP 113. Springer, Berlin
- 6th SIGSAND/PLAIS EuroSymposium 2013: Wrycza, S (ed) (2013) Information Systems: Development, Learning, Security, LNBIP 161. Springer, Berlin
- 7th SIGSAND/PLAIS EuroSymposium 2014: Wrycza, S (ed) (2014) Information Systems: Education, Applications, Research, LNBIP 193. Springer, Berlin
- 8th SIGSAND/PLAIS EuroSymposium 2015: Wrycza, S (ed) (2015) Information Systems: Development, Applications, Education, LNBIP 232. Springer, Berlin
- 9th SIGSAND/PLAIS EuroSymposium 2016: Wrycza, S (ed) (2016) Information Systems: Development, Research, Applications, Education, LNBIP 264. Springer, Berlin
- 10th SIGSAND/PLAIS EuroSymposium 2017: Wrycza, S, Maślankowski, J (eds) (2017) Information Systems: Development, Research, Applications, Education, LNBP 300, Springer, Berlin

- 11th SIGSAND/PLAIS EuroSymposium 2018: Wrycza, S, Maślankowski, J (eds) (2018) Information Systems: Research, Development, Applications, Education, LNBIP 333, Springer, Berlin
- 12th SIGSAND/PLAIS EuroSymposium 2019: Wrycza, S, Maślankowski, J (eds) (2019) Information Systems: Research, Development, Applications, Education, LNBIP 359, Springer, Berlin

The 13th EuroSymposium was organized by the Polish Chapter of Association for Information Systems (PLAIS) and the Department of Business Informatics of the University of Gdańsk, Poland.

PLAIS was established in 2006 as the joint initiative of Claudia Loebbecke, former president of AIS, and Stanislaw Wrycza, University of Gdańsk, Poland. PLAIS co-organizes international and domestic IS conferences, and gained the title of Outstanding Chapter of Association for Information Systems in 2014, 2016, 2017, and 2018.

The Department of Business Informatics of the University of Gdańsk conducts intensive teaching and research activities. Some of its academic books are bestsellers in Poland, like the recent "Business Informatics. Theory and Applications" published by PWN in 2019 (in Polish). The department is active internationally. The recent leading areas of the department's research and publishing in international academic IS journals include the unified theory of acceptance and use of technology as well as information technology occupational culture (ITOC). The department is a partner of the European Research Center for Information Systems (ERCIS) consortium and SAP University Alliances. Students of the Department of Business Informatics of the University of Gdańsk have recently been awarded several times for their innovative projects in a yearly AIS Student Chapters Competition in the USA.

EuroSymposium 2021 received 34 papers from authors in 13 countries. The paper submission and reviewing process was supported by the Open Conference System (OCS) hosted by Springer. The members of International Program Committee carefully evaluated the submissions, selecting 10 papers for publication in this Springer LNBIP volume 429 (an acceptance rate of 29%), with submissions divided into the following four sessions in the event program:

- Digital enterprises
- Smart cities
- Digital education
- Innovative methods in data and process analysis

I would like to express my thanks to all authors, reviewers, the Advisory Board, the International Program Committee, and the Organizing Committee for their support, efforts, and time. Along with the participants, they have all contributed to the successful accomplishment of EuroSymposium 2021.

July 2021   Stanisław Wrycza

Organization

General Chair

Stanislaw Wrycza   University of Gdańsk, Poland

Organizers

The Polish Chapter of Association for Information Systems (PLAIS)
Department of Business Informatics at the University of Gdańsk, Poland

Patronage

European Research Center for Information Systems (ERCIS)
Committee on Informatics of the Polish Academy of Sciences

Advisory Board

Wil van der Aalst	RWTH Aachen University, Germany
Joerg Becker	European Research Centre for Information Systems, Germany
Jane Fedorowicz	Bentley University, USA
Dimitris Karagiannis	University of Vienna, Austria
Claudia Loebbecke	University of Cologne, Germany
Wojciech Penczek	Committee on Informatics of the Polish Academy of Sciences, Poland
Keng Siau	Missouri University of Science and Technology, USA

International Program Committee

Jerzy Auksztol	University of Gdańsk, Poland
Akhilesh Bajaj	University of Tulsa, USA
Dmitriy Demin	National Technical University "Kharkiv Polytechnic Institute", Ukraine
Michal Dolezel	Prague University of Economics and Business, Czech Republic
Petr Doucek	Prague University of Economics and Business, Czech Republic
Helena Dudycz	Wrocław University of Economics, Poland
Krzysztof Goczyła	Gdańsk University of Technology, Poland
Marin Fotache	Alexandru Ioan Cuza University of Iaşi, Romania
Xavier Franch	Universitat Politècnica de Catalunya, Spain

Rolf Granow	Luebeck University of Applied Sciences, Germany
Antoine Harfouche	Université Paris Nanterre, France
Ola Hendfridsson	Miami Herbert Business School, USA
Arkadiusz Januszewski	University of Science and Technology in Bydgoszcz, Poland
Piotr Jędrzejowicz	Gdynia Maritime University, Poland
Dorota Jelonek	Częstochowa University of Technology, Poland
Bohdan Jung	Warsaw School of Economics, Poland
Kalinka Kaloyanova	Sofia University, Bulgaria
Iryna Karachun	Belarusian State University, Belarus
Marite Kirikova	Riga Technical University, Latvia
Vitaliy Kobets	Kherson State University, Ukraine
Jolanta Kowal	University of Wrocław, Poland
Henryk Krawczyk	Gdańsk University of Technology, Poland
Tim A. Majchrzak	University of Agder, Norway
Svetlana Maltseva	Higher School of Economics, Russia
Marco de Marco	Università Cattolica del Sacro Cuore, Italy
Ngoc-Thanh Nguyen	Wrocław University of Science and Technology, Poland
Marian Niedźwiedziński	University of Computer Science and Skills in Łódź, Poland
Mieczysław L. Owoc	Wrocław University of Economics, Poland
Nava Pliskin	Ben-Gurion University of the Negev, Israel
Jolita Ralyte	University of Geneva, Switzerland
Vaclav Repa	Prague University of Economics and Business, Czech Republic
Michael Rosemann	Queensland University of Technology, Australia
Thomas Schuster	Pforzheim University, Germany
Marcin Sikorski	Gdańsk University of Technology, Poland
Janice C. Sipior	Villanova University, USA
Piotr Soja	Cracow University of Economics, Poland
Reima Suomi	University of Turku, Finland
Jakub Swacha	University of Szczecin, Poland
Abbas Tarhini	Lebanese American University, Lebanon
Pere Tumbas	University of Novi Sad, Serbia
Eleonora Veglianti	Università Cattolica del Sacro Cuore, Italy
Catalin Vrabie	National University of Political Studies and Public Administration, Romania
Samuel Fosso Wamba	Toulouse Business School, France
Yinglin Wang	Shanghai University of Finance and Economics, China
Janusz Wielki	Technical University of Opole, Poland
Till J. Winkler	Copenhagen Business School, Denmark
Radosław Zajdel	University of Łódź, Poland
Andrew Zaliwski	Whitireia New Zealand, Auckland, New Zealand
Iryna Zolotaryova	Kharkiv National University of Economics, Ukraine
Joze Zupancic	University of Maribor, Slovenia

Organizing Committee

Stanislaw Wrycza (Chair)	University of Gdańsk, Poland
Anna Węsierska	University of Gdańsk, Poland
Jacek Maślankowski	University of Gdańsk, Poland
Dorota Buchnowska	University of Gdańsk, Poland
Bartłomiej Gawin	University of Gdańsk, Poland
Przemysław Jatkiewicz	University of Gdańsk, Poland
Dariusz Kralewski	University of Gdańsk, Poland
Patrycja Krauze-Maślankowska	University of Gdańsk, Poland
Michał Kuciapski	University of Gdańsk, Poland
Bartosz Marcinkowski	University of Gdańsk, Poland
Monika Woźniak	University of Gdańsk, Poland

Contents

Innovative Methods in Data and Process Analysis

Digital Enterprises

Digital Maturity of Retail Enterprises in Ukraine: Technology of Definition and Directions of Improvement

Nadiia Proskurnina[1] , Mikolaj Karpinski[2]([✉]) , Olena Rayevnyeva[1] ,
and Roman Kochan[2]

[1] Simon Kuznets Kharkiv National University of Economics, Kharkiv, Ukraine
[2] University of Bielsko-Biala, Bielsko-Biala, Poland
{mkarpinski,rkochan}@ath.bielsko.pl

Abstract. In the domestic retail sector, digital transformation processes have proceeded rather slowly, therefore, in terms of digitalization, there has been a significant lag behind global trends. However, modern conditions and new challenges have become a catalyst for accelerating digital transformation processes in retail. The development of online sales and communication channels and the digitalization of tools for interaction with market entities have become a priority, which necessitates a new practical rationale and development of recommendations for assessing the degree of digital maturity of retail enterprises. The aim of the article is to develop a technology for analyzing the level of digital maturity of a trading enterprise based on the formation of a digital maturity matrix and an assessment of their digital infrastructure, digital skills of personnel and digital activities. In the context of the directions that are being identified, the analysis of the state of digitalization of the domestic economy for 2017–2019 was performed. In order to prove the empirical and methodological hypothesis the technology of determining the level of digital maturity of the retail enterprise is substantiated. This basis is a set of methods and tools that identifies the reasons for the lag of digitalization processes from market needs. During the research in order to visualize data and research results, the method of expert assessments, the method of systematization and the graphical method have been used. Correlation of certain ranks of importance with the level of digital activity of marketing activities of retail enterprises allowed to build a matrix for determining the level of digital maturity, which allows to establish the level and identify the state of digitalization of marketing activities of retail enterprises. According to the results of the assessment of digital maturity, the highest rank of importance has a website as a sales channel and e-commerce, social media as communication channels and availability of high-speed broadband Internet access.

Keywords: Retail trade · Digitalization · Digital maturity · Digital transformation

© Springer Nature Switzerland AG 2021
S. Wrycza and J. Maślankowski (Eds.): PLAIS EuroSymposium 2021, LNBIP 429, pp. 3–21, 2021.
https://doi.org/10.1007/978-3-030-85893-3_1

1   Introduction

The digital transformation of retail covers a huge number of processes, interactions, transactions, technological changes, internal and external factors. Digital innovations - from big data, advanced analytics, artificial intelligence, machine learning and mobility to the Internet of Things - influence consumer behaviour and accelerate other innovations and transformations. In this regard, there is a need to monitor and assess the level of digital maturity, especially the marketing activities of retailers.

The choice of digital technology depends on the financial capabilities of the enterprise, as well as on the ability to create values for the buyer.

Well-known models of digital maturity assess the opportunities and gaps that allow companies to be competitive, helping them identify areas of digital development, critical skills and the most productive tools.

Currently, many models have been developed for assessing the digital maturity of enterprises. Some of them identify three key areas of digital transformation in transforming customer experience, transforming operational processes and business models. Others use a larger number of enlarged areas of assessment such as strategy & governance, products & services, customer management, operations & supply chain, corporate services & control, information technology, workplace & culture. Sometimes there are only 5 assessment areas combined in the model: vision & strategy, digital talent, digital first processes, agile sourcing & technology, governance.

For the domestic economy and retail trade, the technology for assessing digital maturity is in its infancy. In order to determine the areas of improvement of retail trade enterprises in the context of digitalization, in this paper a technology for analyzing the level of digital maturity based on the formation of a digital maturity matrix.

The aim of the article is to develop a technology for analyzing the level of digital maturity of a trading enterprise based on the formation of a digital maturity matrix and an assessment of their digital infrastructure, digital skills of personnel and digital activities.

In order to achieve the aim the following tasks have been set and solved:

1. An expert survey on three selected areas of trade enterprises digitalization: digital infrastructure; digital skills, digital activity has been carried out.
2. The level of weight and ranking according to the degree of indicators significance by the method of expert assessments has been determined.
3. The methods of comparison and benchmarking have been used to analyze digital activity, i.e., the degree of involvement of trade enterprises in the use of digital tools in operational marketing activities.
4. Basing the matrix approach, the level of digital maturity of trade enterprises has been set on a certain scale.

2   Literature Review

The active use of digital technologies and tools by consumers leads to a change in patterns of consumer behaviour. Key (2017) notes that the modern consumer has become an active participant in the market and interaction with it requires serious effort. Its choice ceased to be spontaneous and impulsive. The purchase decision is now formed over a long period of time. Chen (1997), Kraus and Kraus (2018) underline that computers, smartphones, and other technological devices have become a natural part of the retail environment. Investigations of Pizhuk (2018) show that today's customers are no longer satisfied with mono-channel retail stores, but expect a highly integrated shopping experience, where they can combine different channels and use them together.

The Oracle report (2018) systematizes global information and communication technology (ICT) drivers and technology priorities. It has been shown that of the nine common digital transformation initiatives: cybersecurity; digital culture; digitization of the back office; digital skills training; use of data in business; digital workplace strategy; new digital services; omnichannel strategy to attract customers is characterized by the largest lag. While the digitization of back office functions was the most complete.

Biryukov et al. (2015) in their investigations proved that the choice of digital technology depends on the financial capabilities of the enterprise, as well as on the ability to create values for the buyer.

The most systematic views on the processes of digital transformation are presented in the studies of Westerman et al. (2014), Laux et al. (2018), Deloitte (2015), McKinsey (2016) and BCG (2018, 2019), company Cognizant (2015), Zaki and Ismail Abdelaa (2018).

According to Deloitte's vision, digital transformation is a continuous process that is constantly improving, and digital maturity is a moving goal (Deloitte 2015), i.e. it can be defined as a momentary indicator that shows not an absolute but a relative degree of a goal achievement. The well-known consulting group BCG defines digital maturity as a measure of an organization's ability to create value through digital technologies, and is, therefore, a key factor in the success of companies embarking on or at the epicenter of digital transformation (BCG 2018).

According to Deloitte research: companies with a high degree of digital maturity are almost 3 times more likely than less mature companies to report high dynamics of net profit and income, which are significantly higher than the industry average (Deloitte 2015). Modern technologies (FinTech, HRTech, FoodTech, MarTech) act as a driver of growth in various market segments. Leading companies in digital maturity have a competitive advantage in several performance indicators, including revenue growth, time to market, profitability, product quality and customer satisfaction. Thus, they have the financial and organizational capacity to support the implementation of innovations, which greatly complicates the ability of outsiders to maintain a competitive position. Even the potential for a digital breakthrough does not provide leadership, as the digital transformation is not only the introduction of new digital technologies and the development of a wide range of technology-related assets and business opportunities, but also a coordinated teamwork and clear organizational strategy. As a rule, digital maturity is created due to the synergistic effect of hard skills and soft skills, and strong leadership, team relations and digital culture have a certain significance for established leaders.

BCG's annual surveys in various regions of the world show that only 2% of companies are on the "digital maturity" stage. Leading companies use the full range of digital capabilities to be useful to the user at any stage of the purchase process. Digital transformation gives an unconditional competitive advantage and allows to increase the company's revenues by 20% and reduce costs by 30% (BCG Study 2018, 2019).

3 Problem Description and Basic Assumptions

3.1 Problem Description

The existence of digital gaps, such as unequal access of citizens to digital technologies and new opportunities, is a major barrier to the development of e-commerce in Ukraine. Among the problems that hinder the development of the Ukrainian economy as a digital one, experts also identify the inconsistency of relevant legislation to global challenges and opportunities; a low level of coverage of the country's territory by digital infrastructures compared to the EU countries; a lack of separate digital infrastructures (for example, the infrastructure of the Internet of Things, electronic identification and trust, etc.); weak state policy on incentives for the development of innovative economy [22].

The identified gaps between the importance and activity in the use of tools for digitization of marketing activities of retailers will identify the ways to overcome them.

3.2 Basic Assumptions

Assumption 1. The essence of the empirical and methodological hypothesis is to prove the relationship between the effectiveness of marketing activities of a commercial enterprise depending on the degree of its digitization.

Assumption 2. The offered scientific and methodical approach to an estimation of a level of digital maturity of operational marketing activity of the trading enterprise which basis is based on the application of a combination of the rank and comparative analysis of certain groups of indicators, a digital infrastructure, digital skills, digital activities will identify the reasons for the lag of digitization processes from market needs.

4 Model Formulation

In order to determine the importance of the selected indicators of digitalization of marketing activities of retail enterprises, an expert survey was conducted.

In order to determine the areas of improvement of the retail trade in the context of digitalization, we have proposed a scientific and methodological approach to determining the level of their digital maturity, schematically shown in Fig. 1.

At the first stage this technology provides a consistent procedure for determining the indicators of digitalization of the industry by conducting an expert survey on three selected areas of digitalization: digital infrastructure, digital skills, digital activity.

At the second stage, among the selected indicators by the method of expert assessments, the level of weight is determined and ranked according to the degree of significance. At the third stage, the methods of comparison and benchmarking analyze digital activity, i.e., the degree of involvement of enterprises in the use of digital tools in operational marketing activities and using a matrix approach sets the level of digital maturity on the scale from primitiveness to leadership.

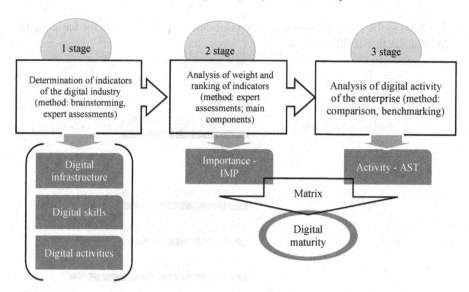

Fig. 1. Technology for determining the level of digital maturity of the retail trade enterprise.

5   Solution Method

5.1   Basis for Method Selection

The systematization of secondary sources of information on the basis of open statistical data allowed to analyze the state of digitalization processes in Ukraine in 2017–2019 in three areas: the degree of digitalization of trade infrastructure; formation of digital skills of the personnel; the level of modern digital tools usage in marketing operations.

The trade infrastructure provides connections and covers all types and forms of relations related to the processes of production, distribution, exchange and consumption, so the level of its development depends on the development of retail trade. A key element of the infrastructure is the ramifications of retail formats: a traditional store trade, an open market trade, i.e., an offline retail; electronic - online commerce and a mixed one. Physical stores are an important part of a customer service in both global and domestic practice, but they are also affected by digital technology. The key players in online retail are online stores, marketplaces, classifies (bulletin boards), price aggregators, companies providing logistics services.

The drivers of transformation are the behavioral factors of the retail market, the growing influence of the buyer, changing needs which encourage the spread of new technologies of trade and interaction. New formats of interaction help to accelerate turnover, increase consumer satisfaction and, ultimately, strengthen the competitiveness of retail businesses.

At the first stage of the analysis, the basic and advanced digital trade infrastructure are distinguished. By a digital trade infrastructure, we mean a set of technologies, products and processes of trade enterprises that provide computing, telecommunications and networking capabilities and operate on a digital (rather than analog) basis [22].

Figure 2 shows the share of domestic enterprises that are provided with the basic digital trade infrastructure in 2017–2019.

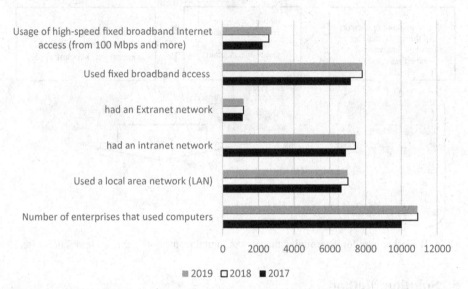

Fig. 2. Dynamics of provision of the trade enterprises with basic digital trade infrastructure in 2017–2019/ Source: systematized on the basis of Statistical Yearbook of Ukraine (2019)

The basic digital trade infrastructure includes computers, fixed broadband Internet, high-speed broadband Internet (from 100 Mbps and more), computer networks Extranet, Intranet, etc. Extranet systems are mostly used by large enterprises. The main function of Extranet-systems is to provide access to formalized information to corporate services, remote divisions of the company, franchise partners, dealer network, wholesale buyers and other partners or customers; it is invisible in search engines. Intranet is an internal corporate network built on the Internet technologies, an intermediate link between the local network and high-level corporate systems - CRM and ERP. From a technical point of view, the Intranet is an internal corporate web-portal created to solve the company's tasks [24].

In 2019, compared to 2018, in trade, as well as in general for all the domestic enterprises, there is a decrease in the level of the use of computers and computer network Intranet (internal corporate web-portal) with increasing the use of high-speed fixed broadband Internet access (from 100 Mbps and more). Thus, the use of computers in Ukraine as a whole decreased by 1.8%, and in trade by 2%, respectively; the use of computers and computer network Intranet (internal corporate web-portal) - by 1.2 and 1.4%; with the growth of broadband Internet access (from 100 Mbps and more) in Ukraine as a whole by 0.9%, particularly in trade - by 0.7%.

The importance of a basic digital trade infrastructure is difficult to overestimate, as it provides the whole process of digitization. The amount of traffic and speed of the Internet connections determine the bandwidth of the digital economy. In particular, the volume of global traffic based on the Internet Protocol (IP), which provides a rough idea of the scale of the data flows, increased from about 100 gigabytes (GB) per day in 1992 to more than 45,000 GB per second in 2017. A data-based economy is only at an early stage of the development; according to forecasts, by 2022 the amount of global IP traffic will reach 150,700 GB per second, which will be the basic infrastructure for the spread of the Internet of Things (UN 2019).

Under the advanced digital infrastructure of trade, we understand the quantitative and qualitative use of the basic digital structure of trade, namely the direction and purpose of use of the Internet by domestic enterprises, directions and purpose of the use of social media by domestic enterprises. Thus, Fig. 3 shows the areas in which the Internet is used, in Fig. 4 there are the purposes of its use.

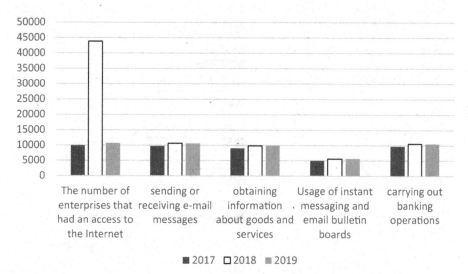

Fig. 3. Directions of the Internet use by the domestic trade enterprises in 2017–2019/ Source: systematized on the basis of Statistical Yearbook of Ukraine (2019)

The active introduction of digital technologies in both business processes and social life of society has led to the growth of almost all indicators of digital activity in Ukraine as a whole and in trade as well. In particular, for trade enterprises the largest growth - 3.6% is stated in the terms of the use of the website for customer service.

In general, in Ukraine, the number of enterprises that have a website (a landing page) is 17,856 units, i.e. 35.2%, among which in trade - 4692, that is 37.6%. The main areas of the use of the website are for customer service; delivery of products (goods, services) online; possibility to order goods and services online; tracking or checking the status of placed orders; personalized content of the website for regular customers; electronic link to the company's profile on social media; vacancy announcements or applications for vacancies online.

The use of websites with multimedia content is even lower, particularly in Ukraine as a whole - by 14.0%, and in trade - by 14.6%. The share of sold products (goods, services) through websites or applications according to the State Statistics Service of

ICT, 2019 in the total turnover in 2019 in the economy as a whole was - 4.5%, in trade - 3.7%.

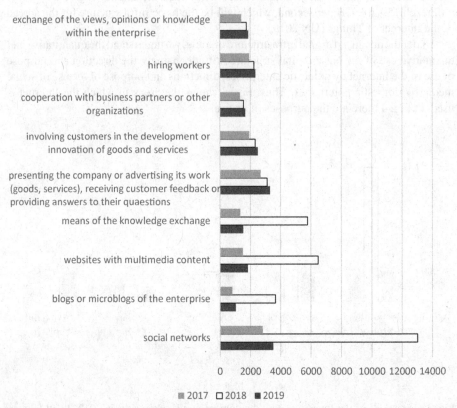

Fig. 4. The purpose of using the Internet by domestic trade enterprises in 2017–2019/ Source: systematized on the basis of Statistical Yearbook of Ukraine (2019)

Retailers use various social media platforms to obtain consumer information. Figure 5 shows the directions and purposes of the use of social media by domestic enterprises, respectively.

The functionality and versatility of social media is an important reason why businesses are increasing their activity on social media. There is an intensification of domestic enterprises and, in particular, those operating in the field of trade in the use of social media. With the help of social media, companies post descriptions of goods or services, reviews of consumers who have already purchased goods, coupons and promotions on social networking sites to increase Internet traffic and influence the customers' behaviour.

In Ukraine as a whole, such activity increased by 1.9%, and in trade by 2.4%, blogs/microblogs - by 0.8% and 1%, respectively.

The share of trade enterprises that use social media to present or advertise the enterprise is only 26.5%, to receive feedback - 20.0%, to attract customers to some development or innovation - 13%.

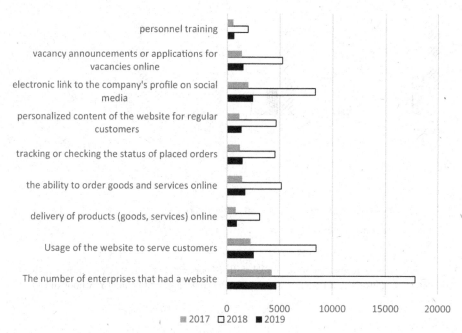

Fig. 5. The purpose of the use of social media by domestic trade enterprises in 2017–2019/ Source: systematized on the basis of Statistical Yearbook of Ukraine (2019)

Customers are a key player in any organization. For the best customer service it is important to know not only the specifics of their behaviour, but also to provide the best experience related, among other things, to the introduction of electronic document management, providing invoices in electronic form. Invoices in electronic form, suitable for the automated processing provided invoices in Ukraine - 19930 companies in total (39.3%), in trade - 5207 units. (41.7%); received invoices for Ukraine - 27097 units (53.5%), in trade - 7065 (56.6%) enterprises. That is, in general, companies are not yet ready for electronic document management which negatively affects the digital development taking into account the growth rate of online sales.

An equally important component of digitalization is the digital skills of staff, as well as the availability of training processes for technical IT professionals, which is the basis of a team integration of enterprises. Because the work of a modern trade enterprise involves the presence of cross-functional teams that bring together specialists in marketing communications, customer analytics, algorithmic advertising, store management, marketing and management. Figure 6 shows data on the formation of different levels of digital skills of personnel of domestic enterprises.

The data of Fig. 6. regarding digital skills in 2018 and 2019, indicate a reduction in basic skills (availability of staff with user skills) and specific in terms of availability and qualification of specialists in the field of ICT both in Ukraine as a whole and in trade. Thus, the availability of personnel with user skills decreased by 1.8% and amounted to 87.9%, and in trade by 2.0% and the scale of 87.3%. The decrease in the availability and qualification of specialists in the field of ICT was 0.7%, and in trade 0.4 in 2019 compared

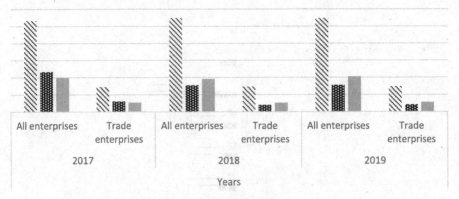

Fig. 6. Indicators of formation of the digital skills of personnel of domestic enterprises in 2017–2019 (share of enterprises, %)/ Source: systematized on the basis of Statistical Yearbook of Ukraine (2019)

to 2018. The intensification of digitalization processes necessitated the emergence of highly qualified specialists, so there is an increase in specific digital skills such as staff training in the field of ICT both in Ukraine as a whole and in trade by 0.4% and 0.7%, respectively, and the number of enterprises that recruited specialists in the field of ICT in trade by 0.1%.

Another important characteristic of digitalization is the use of cloud computing (see Fig. 7). Cloud computing is a way of delivering technological resources to users from remote nodes, a kind of software outsourcing, storage and data processing.

There are three main models of cloud computing depending on the type of resources provided: providing end users with full-featured products - software as a service (SaaS); infrastructure as a service (IaaS) provides system administrators with a secure network and storage capacity; the platform as a service (PaaS) is somewhere in the middle, providing the developers with building blocks for creating applications, freeing them from tedious internal problems (Webicom 2019).

Cloud computing was used by 1,439 trading companies in the following areas: e-mail, office software, enterprise database hosting, file storage service, financial or accounting applications, customer relationship management software, computer power for enterprise software. Separately in cloud technologies the use of CRM-systems deserves attention.

For digital development, the coverage rate of 2.9% of all domestic enterprises and trade enterprises, in particular 3.7%, which is 466 units, is extremely low. 1279 trade enterprises bought cloud computing services, 10.3% of enterprises from general servers of service providers; from servers of service providers reserved exclusively for the enterprise - 2.6%.

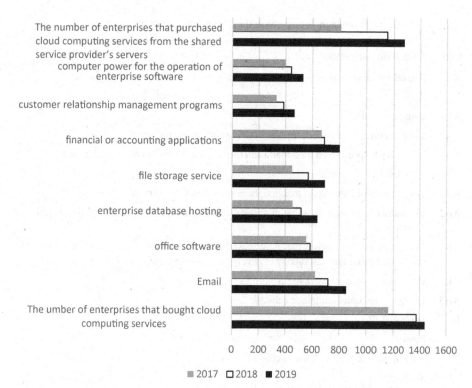

Fig. 7. Areas of the use of cloud computing services by domestic enterprises in 2017–2019/ Source: systematized on the basis of Statistical Yearbook of Ukraine (2019)

The use of an important digital tool - Big Data analysis - is characterized by the same insignificant coverage. Among self-employed enterprises, 10.5% performed the analysis, with the involvement of external suppliers - 4.3%.

5.2 Solution Steps and Procedures

The sequence of implementation, highlighted stages in Fig. 1, involves determining the evaluation parameters. In February 2020, we conducted a study of expert opinions and analysis of empirical data on the digital technologies used in the practice of trade business in 2019. It is established that for the modern development of the retail industry in Ukraine the most popular are the following 22 parameters given in Table 1.

For further analysis, a questionnaire was developed, a pilot study was conducted, its validation was established, and an expert survey was conducted, in which 25 experts took part, 5 of whom were administrative staff of the retail enterprises, 10 people were the heads of the retail departments and/or marketing services trade, 10 people - digital marketing specialists.

Experts were asked to determine the rank of the importance of the selected areas of digitalization of retail enterprises.

Table 1. Parameters for assessing the digital maturity of trade enterprises

Code	Evaluation parameters
A	**Digital infrastructure**
	Basic
AA1	Using computers
AA2	Using a computer network Intranet (internal corporate web-portal)
AA3	Use of high-speed fixed broadband Internet access (from 100 Mbps and more)
	Advanced
AB1	The number of enterprises that bought cloud technology services
AB2	The number of enterprises that purchased cloud computing services from the shared service providers
AB3	The number of enterprises that purchased cloud computing services from the servers of the service providers reserved exclusively for the enterprise
AB4	Use of electronic invoices for customers
AB5	Using the Extranet computer network
B	**Digital skills**
	Basic
BA1	Availability of staff with user skills
	Specific
BB1	Availability and qualification of specialists in the field of ICT
BB2	Staff training in the field of ICT
BB3	The number of enterprises that recruited specialists in the field of ICT
C	**Digital activities**
CA1	Website availability (landing page)
CB1	Using the website to serve customers
CB2	Use of websites with multimedia content
CA2	Use of social networks
CB3	Use of blogs/microblogs
CB4	Social media for presenting or advertising the company
CB5	Social media for feedback
CB6	Social media to engage customers in development or innovation
CB7	CRM - Programs for customer relationship management
CB8	Big Data on your own
CB9	Big Data by external service providers
CB10	Businesses that sold via the Internet

The results of the expert assessment were checked for consistency of experts' opinions. The reliability of the results of expert assessments is established by determining the degree of agreement of expert opinions, by calculating the concordance coefficient. With the absence of related ranks, it is calculated by the following formula [26]:

$$W = \frac{12 \times S}{N^2 \times (M^3 - M)} \tag{1}$$

where S is the total square deviation of the total events from the mean value; M - the number of factors; N - number of experts.

The value of S is calculated by the formula (Krymsky et al. 1990):

$$S = \sum_{j=1}^{n} \left(\sum_{i=1}^{m} X_{ij} - \frac{\sum_{j=1}^{n} \sum_{i=1}^{m} X_{ij}}{2} \right) = \sum_{j=1}^{m} \Delta^2 \tag{2}$$

To confirm the non-randomness of the concordance coefficient, its significance was checked, for which Pearson's criterion (X^2), calculated by the formula (Krymsky et al. 1990) is used:

$$X^2 = \frac{S}{N \times M (M + 1)/12} \tag{3}$$

Pearson's calculation criterion is larger than the tabular one (19.67514), so the concordance coefficient W = 0.801 is not a random value, and, therefore, the obtained results can be used in further studies. According to given in Table 2 scales for determining the consistency of expert opinions, the value of W = 0.801 characterizes the high degree of consistency of expert opinions.

Table 2. Scales for determining the consistency of expert opinions (Krymsky et al. 1990)

The value of the concordance coefficient	Degree of agreement of experts' opinions
0–0,2	Lack of consensus of experts
0,21–0,4	Low degree of agreement of experts
0,41–0,6	The average degree of agreement of experts
0,61–0,8	Quite a high degree of consensus of experts
0,81–1	Absolute consistency

Maturity, in essence, characterizes the level at which the processes of digitalization of marketing, as well as the skills and infrastructure necessary to succeed in the operational activities of retail enterprises. In order to assess the level of maturity, the matrix is shown in Fig. 8.

		Imp – importance		
		Not significant	*Important*	*Very important*
Act – Activity	*Low level*	Digital primitiveness	Digital immaturity	The birth of digital maturity
	Average	Digital backwardness	Formation of digital maturity	Development of digital maturity
	High level	Underdeveloped digital maturity	High digital maturity	Digital leadership

Fig. 8. Matrix for determining the level of digital maturity of enterprises according to the criteria of importance-activity. Source: made by the author

6 Computational Case

The maturity assessment was carried out according to the two-criteria system "importance-activity". The importance of digitization parameters was determined according to the questionnaire of experts. The following scale of importance was used: from 1 to 4 - very important; from 5–8 - important; from 9–12 - non-significant.

The activity of enterprises demonstrates the scale of coverage of enterprises with digital infrastructure, the development of digital skills of the staff and the prevalence in practice of digital marketing tools. The scale of coverage of the enterprises up to 10% indicates a low level of activity; from 11 to 50% - a medium level, and above 50% - a high level.

In order to determine the activity of enterprises in the industry, the indicators of coverage of retail enterprises (the share of enterprises in the total number of retail enterprises) were analyzed:

AA1 - Use of computers; AA2 - Use of computer network Intranet (internal corporate web-portal); AA3 - Use of high-speed fixed broadband Internet access (from 100 Mbps and more); AB1 - The number of enterprises that purchased cloud technology services; AB2 - The number of enterprises that purchased cloud computing services from shared service providers' servers; AB3 - The number of enterprises that purchased cloud computing services from servers of service providers reserved exclusively for the enterprise; AB4 - Use of electronic invoices for customers; AB5 - Using the Extranet computer network.

Also indicators of the formation of digital skills of the staff (the share of enterprises in the total number of retail enterprises):

VA1 - Availability of the staff with user skills; BB1 - Availability and qualification of specialists in the field of ICT; BB2 - Staff training in the field of ICT; BB3 - The number of enterprises that recruited specialists in the field of ICT.

The results are presented in Fig. 9a) and b).

Presented in Fig. 9. profiles show uneven formation of basic digital infrastructure in retail enterprises, as the use of computers and computer network Intranet is at a satisfactory level (more than 60%), but the use of high-speed fixed broadband Internet access - at the initial one, as it covers about 20% of enterprises. All other indicators of the advanced level, such as the use of cloud technology services, remain at a low level which is about 10% of the surveyed enterprises.

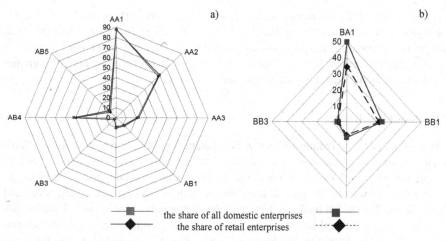

Fig. 9. Profile of digital infrastructure and digital skills of retail enterprises in 2019. Source: developed by the author

In order to determine the prevalence in practice of digital marketing tools, the share of enterprises in the total number of retail enterprises, which have:

CA1 - availability of the website; CB1 - Use of the website for customer service; CB2 - Use of websites with multimedia content; CA2 - Use of social networks; CB3 - Use of blogs/microblogs; CB4 - Social media for presenting or advertising the company; CB5 - Social media for feedback; CB6 - Social media to attract customers to development or innovation; CB7 - Customer Relationship Management (CRM) Programs; CB8 - Big Data on its own; CB9 - Big Data by external service providers; CB10 - Businesses selling via the Internet (excluding orders received by e-mail). The results are presented in Fig. 10.

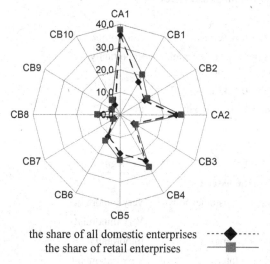

Fig. 10. Profile of digital marketing activities of retail enterprises in 2019/ Source: developed by the author

The profile of digital marketing activities of retail enterprises, shown in Fig. 11, confirms the trend of increasing the potential for the use of basic tools and the weak activity of enterprises in the development of modern digital marketing tools. However, in general, the current state of digitalization of trade enterprises is characterized by greater development than the average for the economy of Ukraine.

6.1 Result Analysis

The results of the evaluation of digitization parameters are shown in summary Table 3. According to the results of the assessment of digital maturity, the highest rank of importance of the success of the operational activities of retail enterprises, according to experts have X9 website as a sales channel and e-commerce X12 (respectively, rank 1 and 2 in Table 3); X10 social media as communication channels (rank 3) and X3 availability of high-speed broadband Internet access (rank 3). The comparison of importance and activity allows to determine the levels of digital maturity according to the proposed matrix (see Fig. 9).

Table 3. Summary table of the results of evaluation of digital maturity parameters

Indexes	Importance	Activity
1. Development of·digital infrastructure:		
1.1. Basic level		
X1 Use of Intranet computer network (internal corporate web-portal)	Importantly	High
X2 Using the Extranet computer network	Not significantly	Low
X3 Use of high-speed fixed broadband Internet access (from 100 Mbps and more)	Very important	Average
1.2. Advanced level		
X4 Use of cloud computing services	Not significantly	Average
X5 Use of electronic document management by enterprises to work with clients	Importantly	High
2. Development of digital skills of staff:		
X6 Availability of staff with user skills	Importantly	High
X7 Availability and qualification of specialists in the field of ICT	Not significantly	Average
X8 Staff training in the field of ICT	Not significantly	Low
3. Development of digital assets:		
X9 Using the Website for Customer Service	Very important	Average
X10 Use of social media	Very important	Average
X11 Using Big Data technology	Importantly	Average
X12 E-commerce via the Internet	Very important	Low

Figure 11 shows the results of testing the proposed scientific and methodological approach. The results clearly demonstrate that the single use of the computer network Extranet (X2) and the underdevelopment of digital staff skills (X8) form a digital primitiveness; digital backwardness is caused by the lack of qualified specialists in the field of ICT (X7); the use of cloud computing services (X4); Big Data technology (X11); the formation of digital maturity is facilitated by the use of basic (X3 - the presence of high-speed fixed broadband Internet access (from 100 Mbps and more) and advanced digital infrastructure (X5 - electronic document management to work with customers), basic staff skills (X6), as well as the use of website for customer service (X9) and social media (X10), the emergence of digital maturity is facilitated by the development of e-commerce (X12).

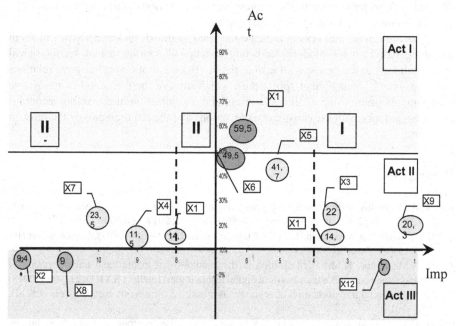

Fig. 11. The results of approbation of the scientific and methodical approach to definition of digital maturity of the enterprises of retail trade/ Source: built by the author on his own research

The study of the best experience of digitalization in domestic retail allows us to say that leaders combine the activity of digital technologies in the field of customer experience (trade and technological process) and in the field of transformation of marketing operations.

7 Conclusion

The analysis allowed to form the author's vision of digitalization areas that will accelerate the development of e-commerce in Ukraine.

The first direction of the digitalization of retail trade is generally connected with the solution of problems of the client's satisfaction with the shopping experience. These include technologies that simplify and speed up the payment process, and various self-service services that eliminate queues, as well as some intelligent loyalty programs that provide maximum personalization of the customer's grocery cart.

The second direction of the digitalization of trade is connected with the use of the format of strategic partnership of trade networks and producers. Strategic partnerships are joint projects to plan production volumes and even to create productions for specific sales volumes, bring new brands to market and intelligent sales management in the interests of customers.

The third direction of the digitalization of trade is the development of distribution infrastructure while maximizing the operational efficiency of all the stages of the product life cycle - from production to the implementation of various digital technologies: the use of chatbots, artificial intelligence capabilities.

These arguments and results indicate the need to transform the activities of retail operators, based on the modern marketing concept of forming a new technological way of thinking and a new way of action, the development of communicative relations with consumers; management approaches, which involve the formation of innovative marketing potential of the retail operator and the digitalization of marketing technologies in accordance with the consumer expectations and the requirements of the modern economy.

References

Key, T.M.: Domains of digital marketing channels in the sharing economy. J. Market. Channels **24**(1–2), 27–38 (2017). https://doi.org/10.1080/1046669X.2017.1346977

Chen, Y.: Marketing innovation. J. Econ. Manage. Strat. **1**, 101–123 (1997). https://doi.org/10.1111/j.1530-9134.2006.00093.x

Kraus, N.M., Kraus, K.M.: Digitalization in the conditions of institutional transformation of economy: basic components and tools of digital technologies. Intellect **XXI**(1), 211–214 (2018)

Pizhuk, O.I.: Bigital transformation of economy as a basis of forming its competitiveness **6**(17), 79–83 (2018)

Oracle: Oracle broadens and deepens its CX appeal with a coherent strategy 2018 (2018). https://www.oracle.com/assets/ovum-cx-appeal-4477523.pdf

Biryukov, D.N., Lomako, A.G., Rostovcev, Y.U.G.: Consideration of anticipatory systems to prevent the risks of cyber threats. Works SPIIRAN **2**(39), 5–25 (2015). https://doi.org/10.15622/sp.39.1

Westerman, G., Bonnet, D., McAfee, A.: The nine elements of digital transformation. MIT Sloan Manage. Rev. **55**(3), 1–6 (2014)

Lauks, D., Makolej, D., Noronha, E., Uejd, M.: Cifrovoj vih': kak pobezhdat' didzhital-novatorov ih zhe oruzhiem. M.: Eksmo (2018)

Deloitte: MIT Sloan Management Review, Strategy, Not Technology, Drives Digital Transformation (2015). https://deloitte.wsj.com/cfo/2015/09/30/strategy-not-technology-drives-digital-transformation/. Accessed 17 Mar 2021

Industry 4.0: Industry 4.0 at McKinsey's Model Factories. Get Ready for the Disruptive Wave. McKinsey Digital (2016). https://capability-center.mckinsey.com/files/downloads/2016/digital4.0modelfactoriesbrochure_0.pdf

BCG Study: Digital Marketing Maturity study 2018, Australia & New Zealand brands (2018). http://think.storage.googleapis.com/docs/BCG-Google-AUNZ-Digital-Marketing-Maturity-Report.pdf

BCG Study: Digital Marketing Maturity study 2019, Asia Pacific brands (2019). https://www.thi nkwithgoogle.com/_qs/documents/9173/Understanding_the_path_to_digital_marketing_mat urity.pdf

DBT: Digital Business Transformation. A Conceptual Framework. 2015 Global Center for Digital Business Transformation (2015). https://ru.scribd.com/document/372049639/Digital-Bus iness-Transformation-Framework-pdf

Zaki, M., Ismail Abdelaa, M.: Digital Business Transformation and Strategy: What Do We Know So Far. Working Paper (2018). https://doi.org/10.13140/RG.2.2.36492.62086. https://www. researchgate.net/publication/322340970_Digital_Business_Transformation_and_Strategy_ What_Do_We_Know_So_Far

Ukraine 2030e - a country with a developed digital economy. https://hvylya.net/analytics/econom ics/ukraina-2030e-kraina-z-rozvinutoju-cifrovoju-ekonomikoju.html. Accessed 15 Jan 2021

UN: Digital economy report 2019 (2019). https://unctad.org/system/files/official-document/der 2019_overview_ru.pdf

Webicom (2019). https://webi.com.ua/ua/article/chto-takoe-ekstranet-i-intranet-sajta. Accessed 01 Feb 2021

Statistical Yearbook of Ukraine (2019). http://www.ukrstat.gov.ua/druk/publicat/kat_u/2020/zb/ 11/zb_yearbook_2019.pdf

Krymsky, S.B., Zhilin, B.B., Paniotto, V.I., et al.: Expert assessments in sociological research. Nauk. Dumka, Kiev (1990). 320 p.

Maturity Models for the Assessment of Artificial Intelligence in Small and Medium-Sized Enterprises

Thomas Schuster, Lukas Waidelich$^{(\boxtimes)}$, and Raphael Volz

Institute of Smart Systems and Services, Pforzheim University of Applied Sciences,
Tiefenbronner Str. 65, 75175 Pforzheim, Germany
{thomas.schuster,lukas.waidelich,raphael.volz}@hs-pforzheim.de

Abstract. The digital transformation creates major challenges for companies and fosters disruptive change processes. Artificial intelligence (AI) and its applications play a major part in this context. Therefore, companies need to assess the necessity and advancement of AI applications on a regular basis. This type of AI assessments of applications, services and products can be driven based on maturity models (MM). This article aims to present and assess the status quo of current research on existing AIMM. Simultaneously, this work defines the foundation for further research activities in the field of AIMM and addresses previously neglected perspectives such as facets of privacy or ethical issues.

Keywords: Artificial intelligence · AI · Maturity level · Maturity model

1 Introduction

A wide range of drivers have fuelled the advancement of artificial intelligence (AI) and its commercial application. Powerful AI solutions are being used in an increasing number of domains, opening great potential. AI applications include the inherent ability to play a critical role in competitive differentiation [1]. Studies indicate that AI is seen as the most important field of action for corporate success. In contrast, there is an inconsistent approach to the conceptual adoption of AI in companies. Especially small and medium-sized enterprises (SMEs) face great challenges [2]. Our work aims to disentangle the currently confused situation of existing AI maturity models (AIMM) and present a solution so that companies can establish a reliable assessment of their AI maturity status. Our research questions for this review paper are therefore:

- Which AIMM are available and what are their focal points?
- How do AIMM help to identify the status in individual environments?
- How do AIMMs help to determine the future steps in AI development?
- In what way could an AIMM that is suitable for SMEs be set up?

© Springer Nature Switzerland AG 2021
S. Wrycza and J. Maślankowski (Eds.): PLAIS EuroSymposium 2021, LNBIP 429, pp. 22–36, 2021.
https://doi.org/10.1007/978-3-030-85893-3_2

In a first step, basics of AI and MM are explained to set the scope for the analysis and the development approach. Then a methodology is outlined as a multi-step process model, followed by a detailed description of the method's implementation. This includes problem definition, comparison of existing AIMM, determination of a development strategy and the iterative maturity model development. Finally, we draw a conclusion and provide thoughts on future research activities.

2 Theoretical Fundamentals

2.1 Artificial Intelligence

Artificial intelligence is up to discussion for a long time already. Alan Turing defined important foundations of AI such as the so-called Turing test [3]. The objective of the test attempts to simulate a condition in which a machine can imitate the thinking of a human being, so that a human subject assumes to interact with another human and not with a machine. Hence, Turing defines a machine to be intelligent if it's the responses are indistinguishable from those of a human [3]. John McCarthy described AI as the science and technology of making intelligent machines, especially intelligent computer programs [4]. Afterwards, numerous AI initiatives and research activities emerged, such that AI experienced its first peak in the 1950s and 1960s [5, 6].

Lack of progress and reduced funding associated with disappointment and disillusionment led to the AI winter in the 1970s [7]. New incentives for AI research were given by the victory of the chess computer Deep Blue against Garri Kasparov in 1997 [6]. From this perspective, it makes sense that AI enjoys increasing popularity from a scientific point of view and serves numerous new research areas and industries such as finance or healthcare or education sector [8, 9]. AI applications are used, for example, in early cancer detection, automated vehicle guidance, or chatbots [6, 10, 11].

Since the emergence of AI as a field of research, risks are heavily discussed as well [12–14]. Amongst other things, this includes human job loss in favour of AI technology, effects of error-prone autonomous machines, or loss of individual privacy.

Nowadays, AI aggregates numerous disciplines such as machine intelligence, computer intelligence, or intelligent behaviour and concepts such as machine learning or deep learning and with the goal of application-oriented AI development, De00, [6, 15–17]. Furthermore, AI can be divided into the types of weak and strong AI. In weak AI, the AI performs a supporting activity in a subdomain, while strong AI assumes independent cognitive behaviour without human intervention. The highest form of AI is artificial superintelligence, which currently only seems possible in theory [6].

2.2 Maturity Models

MM initially targeted to improve software processes in organizations [18]. Since then, MMs have become established across domains and are being applied in many fields, such as information systems (IS), development, sustainability, health or business process management, digital transformation, and Industry 4.0 [19]. In general, MMs are used in innovative, novel, and complex domains that require a systematic and structured approach [20]. AI specific MMs are primarily defined by individuals from the practical rather than the scientific environment.

MMs serve as a management tool to assess the capabilities of an organization [21], create initial awareness of a topic, provide a valid assessment of the status quo by evaluating individual capabilities and may initiate innovation processes [22]. MMs can be descriptive, prescriptive, or comparative. A descriptive MM is used to analyse the current situation and examines existing characteristics according to predefined criteria. A prescriptive MM is used for the target achievement of an existing maturity level or the specification of corresponding improvement measures. A comparative MM enables a benchmark between internal or external entities. MMs can combine the characteristics.

2.3 Small and Medium-Sized Enterprises

SMEs differ from large enterprises in numerous ways. In general, SMEs have a small number of employees. They have a broad range of expertise, but usually do not have an academic background. SMEs are characterised by short, direct information paths, strong personal relationships, few coordination problems, a low degree of formalisation and a high degree of flexibility. SMEs are often run by owners or entrepreneurs. As a rule, they do not make decisions in consultation with others, improvise or rely on their intuition. Rarely is management based on long-term planning. These characteristics are generalised and do not apply to all SMEs [23].

SMEs are therefore not approached by scientifically sophisticated content, but by elements that are easy that can be implemented operationally and quickly lead to initial successes. Scientific work must be transferred to an application-oriented stage [23].

3   Research Methodology

The Design Science Research (DSR) approach according to Hevner et al. [24] is used as research method. The procedure forms the basis for the conception of a design arte-fact and the development of the profound AIMM. Our research is guided towards the seven steps proposed by Becker, Knackstedt and Pöppelbuß [25] (see Fig. 1) for the development of MM and aims to develop a new design artefact in the form of an AIMM.

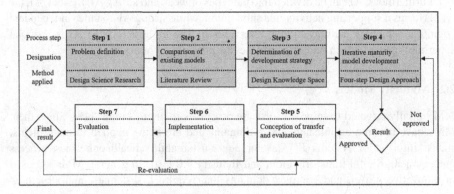

Fig. 1. Procedure model for developing a maturity model based on [25].

This research addresses four of the seven steps of the MM process model: *(1) Problem Definition, (2) Comparison of existing Models, (3) Determination of development strategy, (4) Iterative maturity model development.* The other steps *(5) Conception of transfer and evaluation, (6) Implementation (7) Evaluation* are currently in progress and will be articulated in future research. Figure 1 shows the necessary process steps for the design artefact and indicates the applied methods for each process step.

3.1 Problem Definition

At the beginning of MM development there is an initial definition of the problem [25]. An essential DSR objective is to gain knowledge and understanding that enables the development and implementation of technology-based solutions to unsolved business issues [24]. A problem should be understood as the difference between the actual state and the target state, e.g., in the context of IS.

3.2 Comparison of Existing Maturity Models

Before a comparison of existing AIMM could be carried out, they first must be identified within the second process step [25] (see Fig. 1). For this purpose, a systematic literature review was conducted, which is essentially based on the methodology of Brocke et al. [26] and comprises five phases: (1) the definition of the basic research parameters [27], (2) the definition of the search terms and search term combinations for the database search, (3) the selection of the databases to be searched, (4) the analysis of the search results and (5) the reverse search for further information sources. The basic research parameters are tabulated:

Table 1. Definition of research parameters by taxonomy according to Cooper [27].

Focus	Focus on research results and practice-oriented use cases of AIMM
Goal	Identification of existing, scientifically sound and application oriented AIMM models for enterprises
Perspective	The perspective of the research is neutral
Scope	Selective literature research focusing on the research field
Organization	Conceptually oriented, pattern recognition in relation to AIMM
Target group	Addressing (IS) researchers and practitioners applying AIMM

The keyword definitions used for the research include the terms Artificial Intelligence and Maturity Model or Readiness Model or Capability Model, as well as their combination. The selection of databases for conference and journal papers focused on the search engines and databases Google Scholar, SpringerLink and ACM, whereby IS conferences and journals with the peer review process were considered, which were subsequently expanded to include AI technical reports. Furthermore, the first 50 results per database or search platform of the last ten years were examined, analyzed by title and abstract, and duplicates and irrelevant results were excluded. Furthermore, a forward and backward search of relevant works was carried out to expand the status to include the new works. In total, 15 AIMM approaches were identified that had artefacts of MM. The results are presented in Sect. 4.

3.3 Determine Development Strategy

The selected procedure model [25] foresees a basic strategic decision after the analysis of existing models. The options include designing a new model, improving an existing model, combining several existing models into a new model, transferring structures or applying content from existing models to new domains [25]. In a new development of an AIMM, strengths such as content and/or structure of existing models should be taken into account and weaknesses should be targeted [21].

As a basis for decision-making, the identified AIMMs are to be grouped into the so-called Design Knowledge Map (DSM), which is to be assigned to the DSR domain. Each DSR project has a starting point that builds on existing Design Knowledge (DK). The DSK enables the classification into a nine-field portfolio of the analysed AIMM according to the criteria projectability (into the problem space), fitness (in the solution space) and confidence (in the evaluation) according to the DSR principles [28]. The criteria may in turn have different characteristics (low, medium, high).

3.4 Iterative Maturity Model Development

The iterative MM development phase based on the previous research on existing AIMM, and strategy consideration is the major element of the process model. For this purpose, the authors propose an iterative four-step design process consisting of the sub-steps selecting the design level, selecting the approach, designing the model section, and testing the results [25]. According to this, the design level should be specified first. At this stage, the basic idea of the maturity stages, the structure of the levels and the dimensions (main and if required subcategories) of the maturity model should be specified. The next step provides two design approaches in practice that can be distinguished. The bottom-up approach starts by defining the dimensions and then derives the maturity levels and their descriptions. The top-down approach proceeds vice versa and first defines the maturity levels and then specifies the dimensions and their descriptions [21]. In the third step, a suitable design model selection must be chosen for each abstraction level. A preferred approach is the use of literature analysis, which identifies success factors, comparable models, and evaluation criteria for the maturity model. Explorative methods such as the Delphi method or the use of creativity techniques (brainstorming or clustering) can be utilized. The last step of the iterative model development must be designed according to the chosen approach. The result should be examined for the requirements completeness, consistency, and problem adequacy. With a positive test outcome, the fourth step of the procedure model step is completed [21, 25, 29].

The selection of the model including the dimensions and maturity levels forms the point of view of a subsequent discussion and exchange with the focus group of AI and IT experts in the following fifth step.

4 Analysis and Development of AIMM in SMEs

4.1 Problem Definition

The influence of AI as an economic success factor is acknowledged [1, 30]. According to a recent study, more than two-thirds of German companies see AI as the most important forward-looking technology, eight out of 100 companies are already using AI solutions and about one-third are planning to introduce it within a reasonable time [2]. Nevertheless, the commercialisation of AI is in an early phase. SMEs in particular are reluctant to use AI [31]. The introduction of AI applications is seen as a challenge in organisations. This circumstance can be attributed to the fact that AI is often not yet comprehensively integrated into the business strategy. At the same time, there are often gaps within the company between organisational units that already understand and use AI and those that do not. Currently, AI solutions play a key role in the continuous improvement of products, business processes and models and thus make a valuable contribution to corporate development [20, 21].

An AIMM serves companies to determine their position. Managers can get an overview of the AI progress, thus critically observe the organisation's performance, and finally take optimising measures. For this, the maturity level must be determined [29]. With our AIMM, we aim to enable companies to drive AI commercialisation.

4.2 Comparison of Existing Maturity Models

The research identified 15 different AIMM approaches. The focus is on the three comprehensive and scientifically developed AIMMs. These are compared in Table 2. In addition, there are scientific artefacts that contain elements of an AIMM that will also be investigated. Furthermore, the search revealed numerous AIMMs that have no empirical basis, lack documentation and can rather be understood as consulting offers of AI solutions by companies.

The three holistic and scientifically based AIMMs have been published within the last three years and take a more general AI focus. The AI MMs include both differentiated dimensions and maturity stages, which must be present to be classified as MM.

Literary works containing AIMM artefacts could moreover be identified. These either serve as content-related preliminary work or combine essential elements of an AIMM. The work by Vakkuri et al. [32] sees a concrete need for research and addresses the need for an established AIMM and gives impulses to consider ethical dimensions. In addition, three publications found describe manifold dimensions for potential AIMM, some of which were derived from empirical studies (see Table 3) [33–35]. Other publications mostly deal with five-stage AI maturity phases without referring to a concrete MM or naming concrete AI dimensions [36–38].

Finally, our search revealed further AIMMs in the form of reports that formally fulfil the requirements of an AIMM but are not considered in detail. The approaches by Gartner [40], Intel [41] and Ovum [42] either do not have a scientific basis or have scientific deficiencies (poor documentation and non-reproducible procedure). Another AI study refers to the maturity level of administrations on an international level and will therefore

Table 2. Comparison of the identified AIMM from literature.

Author/s	Alsheibani et al. [29]	Yams et al. [20]	Holmstrom [39]
MM name	Artificial Intelligence Maturity Model	AI Innovation Maturity Index	AI Readiness Framework
Focus	AI general	AI general Innovation management	AI general Industry 4.0
Dimensions	AI functions Data structure People Organizational	Strategy Ecosystem Mindset Organisations Technologies Data	Technologies Activities Boundaries Goals
Maturity stages	Initial Assessing Determined Managed Optimise	Foundational Experimenting Operational Inquiring Integrated	None Low Moderate High Excellent
Purpose	Identification and baseline for improving the AI status in the company	Syst. support for AI integration in innovation systems	Support for the development of AI business ideas

Table 3. Overall comparison of AI dimensions from identified research.

Dimensions	Author/s					
	[29]	[20]	[39]	[35]	[34]	[33]
Activities			X			
Analytics				X		
Data	X	X		X	X	X
Decisions				X		
Ecosystem		X				
Knowledge					X	X
Mindset/culture		X			X	X
Organization	X	X	X	X	X	
Resources/people	X			X	X	X
Risk proclivity					X	
Strategy/goals		X	X	X		X
Technology/function	X	X	X		X	

not be analysed [43]. This benchmark serves as a basis for further AIMM development and is at the same time a prerequisite for determining the development strategy.

Many MMs try to depict the complex issue based on a few AI dimensions (see Tables 2 and 3), thus addressing the topic in a rather generalist manner and adding little value through this superficial view. The complete AIMMs hardly differentiate according to criteria such as sector affiliation or company size. So-called key performance indicators (KPI) for classification into the maturity stages of the respective AI dimensions could not be identified either. We were also unable to identify an AIMM that has been proven in practice and has already successfully passed through an evaluation phase. Furthermore, we miss measures of action for the respective maturity phases, i.e., concrete steps how the evaluated organisation can continuously improve to be able to reach the next maturity level. Finally, a cost-benefit analysis should be made for each AI measure in the company. Not every achievement of the next higher stage of AI maturity is commercially viable, so it should be carefully reviewed and considered in the AI MM.

4.3 Determine Development Strategy

The three identified AIMM from Table 2 are classified into the DKM based on the three criteria projectability of the problem space, fitness of the solution space and confidence in the evaluation. The result is shown in Fig. 2.

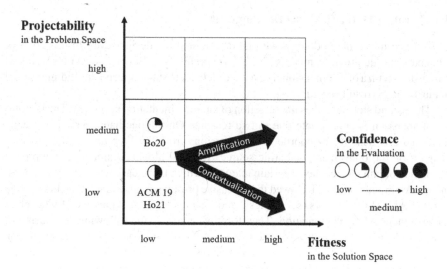

Fig. 2. Classification of the identified preparatory work in DKM and strategic orientation.

As already noted, the DSR area AIMM is still novel, but can already build on basic DK. The two publications by the authors Alsheibani et al. [29] and Holmstrom [39] show the basic knowledge in the problem and solution space mentioned. Confidence in the evaluation has yet to be provided by practice validation, but comparable approaches can be found in the field of AIMM artefacts. In our opinion, the publication by Yams et al.

[20] shows a higher projectability of the problem space. However, this is accompanied by a lower confidence in the evaluation.

For our further work, we follow the approach of improving an existing model by combining several existing models into a new model. The dimensions of the AIMM should be fundamentally reconsidered. From our point of view, the dimensions of data protection and ethics should be considered in an AIMM. New technology increasingly brings new challenges with regard to personal data and pose ethical questions. For this reason, these aspects should be considered and reflected in the AI MM. SMEs in particular are facing major challenges with regard to AI and are in serious danger of losing touch [2]. Here we see great potential for action and the opportunity to close a research gap by further developing the current state of AIMM with an SME focus. Two central strategies with the thrust of increasing the suitability of the solution space come into question here (see Fig. 2). Contextualisation reduces the projectability of the problem space by only considering a specific problem. An example could be a specifically targeted AIMM for SMEs or an industry focus. Amplification aims to maintain the level of projectability of the problem space and increase the fitness of the solution space. This is associated with a higher effort compared to contextualisation. A cross-industry AIMM can be mentioned as an example. The strategy of contextualisation to SMEs makes sense in this case. In addition, an AIMM suitable for SMEs could be designed with corresponding recommendations for action and conclusive explanations.

4.4 Iterative Maturity Model Development

As outlined in the methodology, we first determined the design level. In this context, adapted from the previous models [20, 29] we prefer the following concept: several AI dimensions with a one-dimensional sequence of discrete stages meets a multidimensional maturity stage (result see Table 5).

The second step describes the selection of an implementation approach. The bottom-down approach is appropriate due to two reasons: On the one hand, we have already identified and described important AI dimensions from the literature in the previous chapter; on the other hand, scientific sources [21, 25] also recommend this approach when the domain has reached a certain maturity, which is the case with AI. In determining the AI dimensions, we followed the extensive prior work Building on the knowledge of other AIMM researchers (see Tables 2 and 3), we address the question of what challenges hinder AI implementation. Therefore, we consider the following structuring of AI dimensions into the categories useful: Culture/Mindset, Data, Ethics, Organization, Privacy, Strategy and Technology. The dimensions, their conceptual definition, a purpose for their selection, and a corresponding reference to the literature are presented in Table 4.

The AI dimension identification of the previous described values is followed by the definition of AI maturity stages. Here, we also rely on a five-level maturity stages that has proven itself in practice, e.g., Capability Maturity Model Integration (CMMI):

Table 4. Relevant AI dimensions for SME focused AIMM.

Dimension	Definition	Purpose	Reference
Culture/mindset	Embraces a workplace culture of innovation and transformation	The workforce needs an AI-innovative environment	[20, 33, 34]
Data	Relate to quantity (amount) as well as quality (structures)	Foundation to implement AI applications	[20, 29, 33–35]
Ethics	Use of AI applications under the premise of responsibility, transparency, fairness, safety and security	Compliance and social ownership	[8, 44, 45]
Organization	Structures and (financial and human) resources that enable the use of AI	Provides a framework for the organization	[20, 29, 34, 35, 39]
Privacy	Compliant handling of sensitive data	Protection of business secrets and personal data	[45]
Strategy	Intentions and plans to advance the commercialisation of AI	Strategic alignment of AI as a success factor	[20, 33, 35, 39]
Technology	Refers to areas of application (tools and technologies)	AI share in processes, products, and services	[20, 29, 34, 39]

Level I: Novice

No strategic focus on AI, no AI-friendly working environment and organisation, is not pursuing AI purposes in data collection and structuring, is still not using AI applications and has not yet developed data protection and ethical guidelines for the use of AI.

Level II: Explorer

The company is developing an AI strategy, wants to create a friendly AI working environment and organisational framework, is developing criteria for AI data use, introduces first piloted AI applications and is preparing AI compliant concepts for data protection and ethical requirements.

Level III: User

The Business has defined the AI vision, created initial awareness for a friendly AI working environment and aligned initial structures and resources with AI, prototypically applies the proposed data structure, deploys AI applications and slowly establishes the data protection requirements and ethical issues in individual cases.

Level IV: Translator
The enterprise has defined a clear AI strategy, the working culture enables AI innovations and AI projects have access to established structures and resources, data is systematically collected for AI purposes and enables the use of numerous AI applications, data protection principles and ethical regulations are followed all over the place.

Level V: Pioneer
The AI corporate strategy is perceived as leading in industry, the prevailing culture drives AI innovations and the organization structure is optimized for AI projects, the AI optimised data collection and structuring enables the standardized use of AI applications across companies based on a fully compliant application of data protection principles and an ethical code of conduct.

Table 5. Applied maturity stages and dimensions in SME focused AIMM.

	Level I Novice	Level II Explorer	Level III User	Level IV Translator	Level V Pioneer
Culture/mindset	No AI-friendly culture	Workforce discovers the benefits of AI	Evidence of an AI-friendly culture	Culture enables AI innovations	Employees boost AI innovation
Data	No criteria for collection and structuring of data	Criteria for data infrastructure defined	Prototypical implementation of the data requirements	Data is largely collected and structured	Data collection and structuring optimised
Ethics	No awareness of AI ethics	AI Ethics policies are evolving	AI ethics guidelines applied in single cases	AI ethics rules are widely established	AI ethics principles are holistically applied
Organization	Structure and resources not aligned with AI	Creation of initial structures and resources for AI projects	Piloted structures and resources enable AI projects	Established structures and resources support AI projects	Structure and resources are optimised for AI projects
Privacy	No awareness of data protection	Data protection is partially considered in AI applications	Privacy is taken into account by the AI teams	Data protection is internalised and widely applied	Data protection is fully integrated and considered

(continued)

Table 5. (*continued*)

	Level I Novice	Level II Explorer	Level III User	Level IV Translator	Level V Pioneer
Strategy	No AI vision and strategy available	Vision and Strategy are pushed internally	Vision is established and actions are defined	Strategy is clearly defined	Strategy is perceived as leading in the industry
Technology	No application of AI Tools	Awareness of AI technology	AI technology is partially used	AI applications are adopted	Use of AI technology is standardised

The third step describes the model selection. Here, the design of an AIMM is created. The selected AI dimensions are compared with the five maturity levels in the form of a matrix. The five maturity levels are used as the basis for the associated AI dimension assessments. In this step, the topics are assigned to the different levels. The structure of our AIMM is visualized in Table 5.

5 Summary and Outlook

Digital transformation and AI are about to become the most important technology drivers in information systems. To that end, companies need to assess their capabilities and necessary next steps in this context, and the need for AI specific maturity models (AIMM) is prevalent. An AIMM must be based on scientific measures and metrics and include a practice-oriented perspective as well. A deep literature review revealed a total of 15 AIMM approaches. However, most of these approaches contain methodological and scientific deficiencies or lack important AIMM elements. Only three models are convincing. On closer examination and classification in sense of the DKM methodology, it can be stated that these cover a relatively small solution space and show deficits in the evaluation. Our work contributes to the assessment of existing AIMM and at the same time identifies valuable potential for further development.

It seems reasonable to further develop the existing AIMM that have been identified as promising. In doing so, the listed points of criticism must be addressed and aspects of data privacy as well as ethical issues must be incorporated. In a next step, we are pushing for the prolongation of the AIMM development process according [25] by entering development phases five to seven. Concurrently we will adhere to the contextualisation strategy [28]. Particularly in SME-friendly AIMM, we see opportunities to provide added value for the penetration of AI applications in SMEs.

Our preliminary AIMM model comprises a total of seven dimensions and 5 levels of maturity. A detailed evaluation with the target group (SME stakeholder) is essential. Based on this, concrete recommendations for action can be derived and help to manage the correct application of the AIMM. Another important step is to develop a concrete and easy to use methodology for AIMM assessment so that companies can measure current state und developments. In addition, automated generation of next steps is an interesting question extending the AIMM development. In fact, AI itself could act as co-pilot to suggest next steps, and companies could automatically receive advise if it is necessary to achieve a next level of AIMM (on company, department, or application level).

References

1. Knapp, P., Wagner, C.: Künstliche Intelligenz schafft neue Geschäftsmodelle im Mittelstand. In: Buxmann, P., Schmidt, H. (eds.) Künstliche Intelligenz, pp. 161–172. Springer, Heidelberg (2019). https://doi.org/10.1007/978-3-662-57568-0_10
2. Bitkom: Künstliche Intelligenz kommt in Unternehmen allmählich voran. https://www.bit kom.org/Presse/Presseinformation/Kuenstliche-Intelligenz-kommt-in-Unternehmen-allmae hlich-voran
3. Turing, A.M.: Computing machinery and intelligence-AM turing. Mind **59**, 433–460 (1950)
4. McCarthy, J., Minsky, M., Rochester, N., Shannon, C.: In: Dartmouth Summer Research Conference on Artificial Intelligence (1956)
5. Fang, J., Su, H., Xiao, Y.: Will artificial intelligence surpass human intelligence? SSRN J. (2018). https://doi.org/10.2139/ssrn.3173876
6. Richter, A., Gačić, T., Kölmel, B., Waidelich, L.: Künstliche Intelligenz und potenzielle Anwendungsfelder im Marketing. In: Deutscher Dialogmarketing Verband (ed.) Dialogmarketing Perspektiven 2018/2019, pp. 31–52. Springer Fachmedien Wiesbaden, Wiesbaden (2019). https://doi.org/10.1007/978-3-658-25583-1_2
7. Makridakis, S.: The forthcoming atificial intelligence (AI) revolution: its impact on society and firms. Futures **90**, 46–60 (2017). https://doi.org/10.1016/j.futures.2017.03.006
8. Carr, S.: "AI gone mental": engagement and ethics in data-driven technology for mental health. J. Ment. Health (Abingdon, England) **29**, 125–130 (2020). https://doi.org/10.1080/09638237.2020.1714011
9. Lee, I., Ali, S., Zhang, H., DiPaola, D., Breazeal, C.: Developing Middle school students' AI literacy. In: Sherriff, M., Merkle, L.D., Cutter, P., Monge, A., Sheard, J. (eds.) Proceedings of the 52nd ACM Technical Symposium on Computer Science Education, pp. 191–197. ACM, New York (2021). https://doi.org/10.1145/3408877.3432513
10. Tschandl, P., et al.: Human–computer collaboration for skin cancer recognition. Nat. Med. **26**, 1229–1234 (2020). https://doi.org/10.1038/s41591-020-0942-0
11. Shen, X., et al.: AI-assisted network-slicing based next-generation wireless networks. IEEE Open J. Veh. Technol. **1**, 45–66 (2020). https://doi.org/10.1109/ojvt.2020.2965100
12. Brey, P.: Freedom and privacy in ambient intelligence. Ethics Inf. Technol. **7**, 157–166 (2005). https://doi.org/10.1007/s10676-006-0005-3
13. Rotman, D.: How technology is destroying jobs. Technol. Rev. **16**, 28–35 (2013)
14. Eidenmueller, H.G.M.: Machine performance and human failure: how shall we regulate autonomous machines? SSRN J. (2019). https://doi.org/10.2139/ssrn.3414602
15. Barr, A., Feigenbaum, E., Roads, C.: The handbook of artificial intelligence. JSTOR **1**, 78 (1982)
16. Schalkoff, R.: Artificial Intelligence: An Engineering Approach. McGraw-Hill, Inc., Singapore (1990)
17. de Silva, C.: Intelligent Machines: Myths and Realities. CRC Press, London (2000)
18. Paulk, M.: A history of the capability maturity model for software. ASQ Softw. Qual. Prof. **12**, 5–19 (2009)
19. Dahlin, G.: What can we learn from process maturity models – a literature review of models addressing process maturity. IJPMB **10**, 495 (2020). https://doi.org/10.1504/IJPMB.2020.110285

20. Yams, N.B., Richardson, V., Shubina, G.E., Albrecht, S., Gillblad, D.: Integrated AI and innovation management: the beginning of a beautiful friendship. TIM Rev. **10**, 5–18 (2020). https://doi.org/10.22215/timreview/1399
21. de Bruin, T., Rosemann, M., Freeze, R., Kulkarni, U.: Understanding the main phases of developing a maturity assessment model. In: 16th Australasian Conference on Information Systems, vol. 8–19 (2005)
22. Wendler, R.: The maturity of maturity model research: a systematic mapping study. Inf. Softw. Technol. **54**, 1317–1339 (2012). https://doi.org/10.1016/j.infsof.2012.07.007
23. Immerschitt, W., Stumpf M.: Merkmale von Klein- und Mittelunternehmen Employer Branding für KMU, pp. 17–33. Springer Gabler, Wiesbaden (2014). https://doi.org/10.1007/978-3-658-01204-5_2
24. Hevner, M.: Park, ram: design science in information systems research. MIS Q. **28**, 75 (2004). https://doi.org/10.2307/25148625
25. Becker, J., Knackstedt, R., Pöppelbuß, J.: Developing maturity models for IT management. Bus. Inf. Syst. Eng. **1**, 213–222 (2009). https://doi.org/10.1007/s12599-009-0044-5
26. Brocke, J., Simons, A., Niehaves, B., Reimer, K., Plattfaut, R., Cleven, A.: Reconstructing the giant: on the importance of rigour in documenting the literature search process. In: Proceedings of the European Conference on Information Systems, vol. 161 (2009)
27. Cooper, H.M.: Organizing knowledge syntheses: a taxonomy of literature reviews. Knowl. Soc. **1**, 104–126 (1988). https://doi.org/10.1007/BF03177550
28. vom Brocke, J., Winter, R., Hevner, A., Maedche, A.: Special issue editorial –accumulation and evolution of design knowledge in design science research: a journey through time and space. JAIS **21**, 520–544 (2020). https://doi.org/10.17705/1jais.00611
29. Alsheibani, S., Cheung, Y., Messom, C.: Towards an artificial intelligence maturity model: from science fiction to business facts. In: Proceedings of the 23rd Pacific Asia Conference on Information Systems (PACIS), vol. 46 (2019)
30. World Intellectual Property Organization: WIPO Technology Trends 2021. Assistive Technology (2021)
31. Bérubé, M., Giannelia, T., Vial, G.: Barriers to the Implementation of AI in Organizations: Findings from a Delphi Study. University of Hawai'i at Manoa Hamilton Library, Honolulu (2021)
32. Vakkuri, V., et al.: Time for AI (Ethics) Maturity Model Is Now (2021)
33. Jöhnk, J., Weißert, M., Wyrtki, K.: Ready or not, AI comes—an interview study of organizational AI readiness factors. Bus. Inf. Syst. Eng. **63**(1), 5–20 (2020). https://doi.org/10.1007/s12599-020-00676-7
34. Mikalef, P., Gupta, M.: Artificial intelligence capability: conceptualization, measurement calibration, and empirical study on its impact on organizational creativity and firm performance. Inf. Manag. **58**, 103434 (2021). https://doi.org/10.1016/j.im.2021.103434
35. Gentsch, P.: AI in marketing, Sales and Service. How Marketers Without a Data Science Degree Can Use AI, Big Data And Bots. Palgrave Macmillan, Cham (2019)
36. Haefner, N., Wincent, J., Parida, V., Gassmann, O.: Artificial intelligence and innovation management: a review, framework, and research agenda☆. Technol. Forecast. Soc. Chang. **162**, 120392 (2021). https://doi.org/10.1016/j.techfore.2020.120392
37. Niewiadomski, P., Stachowiak, A., Pawlak, N.: Knowledge on IT tools based on AI maturity – industry 4.0 perspective. Procedia Manuf. **39**, 574–582 (2019). https://doi.org/10.1016/j.promfg.2020.01.421
38. Ellefsen, A.P.T., Oleśków-Szłapka, J., Pawłowski, G., Toboła, A.: Striving for excellence in AI implementation: AI maturity model framework and preliminary research results. LogForum **15**, 363–376 (2019). https://doi.org/10.17270/J.LOG.2019.354
39. Holmstrom, J.: From AI to digital transformation: the AI readiness framework. Business Horizons (2021). https://doi.org/10.1016/j.bushor.2021.03.006

40. Sicular, S., Elliot, B., Andrews, W., Hamer, P.: Artificial Intelligence Maturity Model (2020)
41. Intel, Data Center Artificial Intelligence: The AI Readiness Model (2018)
42. Pringle, T., Zoller, E.: How to Achieve AI Maturity and Why It Matters (2018)
43. Shearer, E., Stirling, R., Pasquarelli, W.: Government AI Readiness Index 2020 (2020)
44. Siau, K., Wang, W.: Artificial intelligence (AI) ethics. J Database Manag. **31**, 74–87 (2020). https://doi.org/10.4018/JDM.2020040105
45. Stahl, B.C., Wright, D.: Ethics and privacy in AI and big data: implementing responsible research and innovation. IEEE Secur. Privacy **16**, 26–33 (2018). https://doi.org/10.1109/MSP.2018.2701164

Experience from Attempts to Implement an e-CRM System in an e-Commerce Micro Enterprise

Arkadiusz Januszewski[(✉)] and Kinga Krupcała

Department of Business Informatics and Controlling, UTP University of Science and Technology, Bydgoszcz, Poland
{Arkadiusz.Januszewski,Kinga.Krupcala}@utp.edu.pl

Abstract. The main goal of the study was to explore the process of e-CRM implementation in order to indicate the basic mistakes the e-commerce micro enterprise made, which led to the failure of the e-CRM system implementation. A literature review provides the background for presenting the results of own research performed for an e-commerce micro enterprise. The empirical research involved a case study with a direct unstructured interview with the company's CEO and Digital manager. The research has shown that the investigated micro enterprise made its mistakes twice and took shortcuts, assuming that it was enough to install the software and start using it for the CRM system to function perfectly in the organization. A detailed pre-implementation analysis was not performed either. The implementation conditions and all its costs were not identified, which is indicated in the literature as one of the critical success factors in IT implementations in small business. Neither did the company foresee the technical problems of the demo version of the software and the hidden initial costs, which turned out to be extremely high. There was also a lack of good communication between the implementation company and the employees, which is considered one of the success factors of an IT project.

The main conclusion is that the company cannot learn from its own mistakes; neither does it understand what mistakes have been made. At the same time, it expects a wide functionality supporting, however, relatively complex customer service processes and the full intuitiveness of the system.

The study also provided detailed information about the desired features of e-CRM software for an e-commerce micro enterprise and the desired functionality.

Keywords: e-CRM · Micro enterprise · e-commerce · Critical success factors

1 Introduction

Digital technologies can help Small- and Medium-sized Enterprises (SMEs) overcome the disadvantage of size [5]. The use of Information-Communication Technologies (ICT) can contribute to the innovation of the business model by creating new distribution channels and ensuring efficient customer service [37].

© Springer Nature Switzerland AG 2021
S. Wrycza and J. Maślankowski (Eds.): PLAIS EuroSymposium 2021, LNBIP 429, pp. 37–50, 2021.
https://doi.org/10.1007/978-3-030-85893-3_3

In the applicable literature, one can find many articles describing the research about ICT in SMEs. However, only few of them concern micro enterprises [2, 8, 14, 16, 28]. Micro enterprises may gain significant benefits from ICT in terms of understanding customers better and developing closer relationships, building upon small enterprise's flexibility and informality [41]. However, they face specific challenges while attempting ICT implementation [42]. Micro enterprises often have little or no ICT training and they are not aware of ICT benefits [27].

The authors claim that it should be different in the e-commerce sector, where many micro enterprises operate, and where IT is crucial for running a business. For those companies, ICT is not an alternative but an imperative for future business success [15]. Electronic mail and the Internet as tools with increased affordability and availability used to be of great importance for micro enterprises [18]. Today, however, to communicate with customers, they require more sophisticated tools than email and websites. Such tools include, e.g., e-CRM (electronic Customer Relation Management) systems. They are web-based solutions for synchronizing customer relationship across communication channels, business functions and audiences and they are seen as a consolidation of traditional CRM with e-business marketplace [30]. The e-CRM is applied to identify, attract, and to develop customer value over time on the Internet [33].

So far, a big study on the CRM in SMEs has been conducted in Poland [13]. The results showed that *"as the companies do not see the potential of new technologies and see no need in implementing ICT innovations, they do not require a high level of digital competences from their employees, and they neither want to pay for them nor invest in them. They also have little knowledge about using cloud computing and are afraid of trying such solutions."*

To the best of the Authors' knowledge, no research has been performed in Poland on the implementation of e-CRM systems in micro enterprises or in the e-commerce sector. The aim of this paper is to share the experience from an attempt to implement e-CRM in an e-commerce micro enterprise in the light of the implementation critical success factors (CSF). When introducing the e-CRM system, the enterprise made many mistakes, and the implementation have been unsuccessful. This study seeks to answer the following research questions: 1) What mistakes were made? 2) Can the company learn from its own mistakes by making further attempts to implement the e-CRM system? 3) To what extent are the CSF and errors committed by the micro enterprise consistent with those described in the literature.

The rest of the paper is structured as follows. In the next section we present the general findings from the literature review in terms of the most common mistakes of CRM implementation, IT in micro enterprises, the importance of CRM delivered in cloud computing (CC) for micro enterprises. In the next points our research goals, method and results are presented and discussed against the background of previous research. Some general case study conclusions are given at the end of the paper.

2   Literature Review

2.1   Critical Success Factors and Most Common Failures in CRM Implementation

In the applicable literature, there is a relatively consistent belief that CRM is a comprehensive solution which includes 3 major components: Technology, People and Business Processes. Most of the research on CSF and on the mistakes made during CRM implementation addresses the three aspects. A very detailed review was made by Almotairi [3], who, from several dozen CRM success factors described in the literature, finally selected the 10 most important ones, guided by the degree of acceptance for the factor by the literature which will be reflected in the percentage of the factor occurrence in the literature and the linking between the success factor and the CRM failure causes: Top Management Commitment (80% of appearances in the literature), CRM strategy (47%), Data management (40%), Culture change (47%), Process change/structure redesign (27%), Management and integration of IT systems (67%), Skilful, motivated and trained staff (40%), Customer involvement/consultation (27%), Monitoring, controlling, measuring and feedback (33%) and Inter-departmental integration (33%). Subsequent studies usually confirmed the important role of those factors [1, 20, 32].

The most common CRM implementation mistakes cover a lack of factors such as top management support, linking CRM project to strategy, a customer-centric culture, achieving success early in the project, change management, involving the final user in designing CRM solutions, as well as thinking of CRM as a pure technology, failing in re-engineering business processes [10, 29, 38].

2.2   Introducing CRM in e-Commerce Micro Enterprises

For many years the dominant view in the literature was that small enterprises have more obstacles in the implementation of ICT than larger enterprises. Small businesses usually endure more restrictions, such as a lack of resources, financial constraints, a lack of experts, and management with short-term insight and so they find it more difficult [4, 6, 12, 42]. The employees of small organizations usually lack professional IT knowledge and skills [34] and so, in terms of reporting and analysing information, large organizations have more potential than the small ones.

The emergence of the Internet allowed smaller organisations to use simple technologies such as websites, email packages and databases [9, 21]. However, to successfully compete with larger companies, they must use more advanced solutions ensuring a high level of customer service. Customer influence is an integral element of SME decision-making in terms of introducing ICT [19]).

A growing number of small entities in the e-commerce sector triggered a high demand for customized e-CRM systems, and the other way around; the increasing availability of those systems created a potential of growth and innovation in their customer-related activities and processes [44, 46]. The studies by Lecerf [33] imply that IT offers SMEs a competitive advantage to stand out on multiple markets and that e-CRM enhances the quality and timing of shipments, decreases costs and optimizes customer satisfaction, as well as facilitates coordinating business processes, such as marketing, sales, and customer services. The research by Desouza and Awazu [17] showed that SMEs implementing IT are more efficient in competing with large international companies.

Software-as-a-Service (SaaS) IT delivery model and cloud computing have created another chance to overcome difficulties in implementing ICT in micro enterprises. SaaS-CRM are cheaper and less complicated to implement, as compared with CRM on-premise and they are, therefore, a more desirable option for SMEs [7, 31] and, for that reason, they should be even cheaper and less complicated for micro enterprises. In SaaS model, CRM applications are installed on servers and users access its functions on the Internet and with a web browser. CRM in the cloud are modern systems fully 'web-enabled' and providing a 'web portal' capability, constituting the web front end to the order-capture function [44].

3   Research Methodology

The subject matter of the article and of empirical research has not been studied in detail so far. The research on CSF in CRM implementation focused on SMEs or on all companies in general. No studies were performed on CRM implementation problems in micro enterprises, especially in the e-commerce sector. The Authors' experience shows that they can be specific to micro enterprises in that sector and quite different from the ones characteristic for larger organizations. To confirm these beliefs, first of all, a single e-commerce micro enterprise with two unsuccessful attempts at implementing the e-CRM system was considered.

3.1   Research Goals

The general goal of the empirical study was to explore the process of CRM implementation in order to indicate the basic mistakes the enterprise made, which led to the failure of the e-CRM system implementation. The achievement of the general goal was to be ensured by the implementation of the following specific goals:

1) Understanding the rationale for implementing the e-CRM system.
2) Knowing the pitfalls and mistakes made during the implementation.
3) Indicating the important-for-e-commerce-micro-enterprise CRM software features and functionalities, the lack of which was one of the opting-out reasons.

3.2   Research Method

The research involved a case study approach with a direct interview method and a free-form questionnaire for data collection. The case study facilitates an in-depth examination of the material [39], it helps the researcher to study the problem in real-life environment [45] and it provides abundant contextual data [35]. Such methodology is recommended if there is no prior knowledge of the research area [22], however, it is also invaluable when we can compare a new case study with the prior knowledge [45].

The context of our research is partially embedded in the lack of prior knowledge about e-CRM implementations in micro enterprises in the e-commerce sector, however, it also refers to the previous knowledge about CRM systems implementation failures in SMEs. With the method used, it was possible to collect detailed qualitative and longitudinal data, which made it possible to delve into aspects of the issues studied.

The interview questionnaire framework was developed from a review of earlier literature studies and included over 30 questions in 2 groups. The first group concerned the premises for implementing the e-CRM system and the implementation process, especially the system testing. The second-group questions concerned the reasons for giving up the system and included, e.g., human aspects, such as management support.

The research was performed in a micro enterprise from the e-commerce sector. The company sells lifestyle devices on the Internet and serves several thousand customers a month. The company sells lifestyle devices on the Internet and serves several thousand customers a month. Its marketing strategy is based on influencer marketing and advertising in social media such as Facebook, Instagram (sponsored posts). The main tools and IT systems currently used in the company are e-mail, communicators on mobile phones and a system for invoicing, handling warehouses and shipments operating in the cloud. Customer service is provided on the Shoplo and Baseline platforms as well as during telephone calls. Google sheets and Dropbox are also used to transfer data between employees.

The study included two interviews: the first one with the company's CEO and the second one with the Digital manager, appointed to implement the e-CRM system. The Digital manager in the organization is responsible for website editing, graphic product design as well as for posts and advertisements, sending technical information and IT service in the company. The interviews were made by the co-author of the study in the first half of April 2021 on the Microsoft Teams platform.

4 Research Results

4.1 The First Attempt to Implement the e-CRM System

The enterprise made an unsuccessful attempt at implementing the e-CRM system two years earlier [25]. The selected free-of-charge Bitrix24 system did not meet the company's expectations. The study showed that the main reasons for abandoning the system were a lack of training, which caused some misunderstanding of the principles of the application operation, a failure to use the functions as well as a lack of application operation skills. Other important reasons were difficulties with switching between the web application and the communicator and a non-intuitive operation. During the previous interview (October 2020), the company was already testing another system, Team Flow, which, in the opinion of the CEO, was to meet the company's expectations due to enhanced functionality, high intuitiveness and due to training and post-implementation support by the provider. Table 1 compares the CEO's opinions on the possibilities of Team Flow presented in the first interview, in October 2020, and in the second interview, in April 2021. The comparison of the results of the interviews demonstrates that the initial impressions from testing the system and contacts with the provider were almost completely wrong.

Table 1. Comparison of TeamFlow's capabilities from two interviews with the CEO

TeamFlow capabilities	The first interview	The second interview
Possibility to implement dedicated functionality	It is possible	It is very expensive
Preparation of personalized application forms	It is possible	It is possible, however, for an additional fee
Possibility of integration with the warehouse system and courier shipments	Such a declaration was made	It never came to the stage when those functionalities were designed
Intuitive operation	Definitely better than in Bitrix24	Very bad
Interface	Better than in Bitrix24	Unfriendly due to the sticky notes system, which makes operation difficult
Phone app	It will be available and it will replace others	Phone app has not been developed and the communicator is missing

4.2 The Second Attempt at Implementing the e-CRM System

The reasons for implementing the second e-CRM (TeamFlow) system turned out almost identical as with the first attempt with the Bitrix24 system: enhanced communication and information flow among the employees and an enhanced service request processing. Additionally, the Digital manager indicated a necessity of having a module for generating invoices for influencers.

The criteria followed while selecting the system again were also similar. The idea was to find a Polish system with a Polish support, operating in CC and offered in the SaaS model to avoid high costs of the initial investment, as it was the case with on-premise systems. The indispensable software features were to include a mobile app and an intuitive interface. The provider selected and, at the same time, the software development company offered a free-of-charge demo to be tested by the enterprise.

As for the first implementation attempt, there was installed a full free-of-charge version of the Bitrix24 system and, more or less after a month, it was no longer used. The second attempt finished after a month of demo testing by a few employees of the enterprise. Prior to testing, the person appointed to implement the system, the Digital manager, was trained by the system provider and, for a few months, the person was testing a full version of the system. His job was to train the employees.

The interviews show that the CEO and the Digital manager gave different answers to the same questions (Table 2).

The Digital manager provided far more detailed answers. First of all, he indicated that the demo version differed considerably from the full version of the system he tested himself. The demo was identified with many errors, which made it very difficult and which, finally, made the system to be given up. The full-version screenshots sent by the

Table 2. Implementation process according to the CEO and the Digital manager

Issue	CEO	Digital manager
System implementation time	A turn of October/November 2020	02.2020–01.2021
Team's system testing time	About one month	About one month (September/October 2020)
Course of implementation	The implementation was dealt with by the Digital manager. No external consultant participated	The provider appointed an account manager–sales rep. The testing time was rough and tumble, all the time something was wrong, something had to be waited for. In the summer holiday period there was no contact whatsoever
Training	There was no training; they system was introduced to the employees by the Digital manager	It took place. There were many consulting meetings for the Digital manager. Later the Digital manager was responsible for sharing the knowledge with the employees
Implementation problems	There were none as, eventually, it was not implemented	Very many problems related with the system demo version, costs, additional functionalities
Reconstruction of the processes in the enterprise	No	No but it was the reason why the system was implemented; to make communication and task management more efficient
Availability of the demo version to everyone in the company	No, only the selected ones	Yes, we could set up to 15 accounts, however, not everyone was testing it, only a few Marketing Department people and the Commercial director
Number of testers	3	5
System modules tested	Service module and teamwork module	Task Management System, RMA module of service requests
Approach of the employees	Unwilling	Totally unwilling

provider were to confirm that a given functionality exists, however, they did not make it possible to verify it.

Next to the unfinished demo system, the problem of costs was essential. The cost of the monthly licence was to be PLN 950, which was confirmed and which the enterprise was prepared for. It turned out that an additional cost of PLN 10 000 of the initial fee for starting the system on the provider's servers and a short training must be paid. Also, the fees for providing additional functionalities, e.g., to process the service requests claims using the website (RMA), appeared to be very high, some tens of thousands zloty. The total costs of the initial fee, for all the additional functionalities the company asked for were to amount to PLN 46 000.

4.3 Reasons for Opting Out of the System

The next group of questions addressed the system defects identified, the mistakes made and other reasons for the decision against the purchase of a full software version.

After one month of the demo version testing, the Digital manager asked the employees to give their opinion on a new system version. The following opinions were given: "the system does not make life easier; it makes the employees tired, a calendar is better, it is totally illegible, logic is missing, the system looks like sticky notes on the wall". The employees were also showing that the version to be tested still had many errors: the system was getting stuck and there were some problems to log in. They used the system unwillingly and they claimed that there was no chance for them to get to like it. They reported on the software being hardly intuitive and they said that it was not possible to create the project-task-subtask hierarchy, which makes task management very messy. In fact, everyone claimed that implementing the system with the learning-from-experience method was a very bad idea.

The interview shows that at the end of 2020 the company received a demo upgrade, which was to look identical as the full version of the system, for testing. After the successive system testing, the Digital manager claimed that the initial training he participated in was too general and too short, whereas the "corrected" version still shows many defects.

However, the reasons for opting out of the system and for an unsuccessful implementation were definitely many more. A complete breakdown of what the CEO and the Digital manager thought is provided in Table 3.

The opinions of the two respondents about the mistakes made are especially interesting. They focus only on the testing stage. The respondents do not understand at all that the implementation success is determined by the adequate actions at all the implementation stages, starting from the system and the provider selection. In fact, the opinion of the CEO shows that no mistakes were made and that it was the software which was the cause of the failure. The Digital manager, however, noticed the advantages of the software yet he claimed that, to benefit from it, a longer training (or possibly films, instructions to be downloaded from the system development company's website) or testing the system's full corrected version would be needed. Unlike the CEO, he claimed that the system had abundant functionalities and that it was mostly the internal communicator which was missing.

The Digital manager also noted that the CRM can be used in the enterprise for additional integrations, e.g., with the courier system, the e-payment and invoicing system, which would be a considerable help not only for the organization but also for the client. An integration with a cloud storage was also considered to store large files (e.g., Google Drive, Dropbox) as the possibilities of storing files with TeamFlow turned out extremely expensive.

Table 3. Reasons for implementation failure according to the CEO and the digital manager

Reasons for implementation failure	CEO	Digital manager
The main reason for opting out	Functionality not useful for task management	High costs, system very much unfinished, one year of testing which did not contribute much
System defects	Not transparent enough; too many things on one screen Useful functions missing. No mobile app. Task allocation and grouping not logical	It did not offer a calendar where all the deliveries could be entered, where absences could be entered, where holiday applications could be filed. It was chaotic, unfinished, unclear, not intuitive for the user The internal communicator was missing and so the employees had to use email or whatsapp
System advantages	There were no advantages, and so it was not implemented	It had advantages but they were hard to see after a short testing time. It made task ordering, configuring the schedule for own tasks and working day possible, however, much time was needed to learn it
Mistakes made during the implementation	The mistakes were not part of testing but they were functional deficits	There must have been a mistake made as it was a system which was new for the employees. Theoretically many things were obvious, however still unclear to us and so we kept arranging meetings with the account manager
Training	There was no training at all	Theoretically yes but only for the Digital manager and impossible to be used further as the demo version was totally different
Functions needed	No	Yes, and for that reason it was selected as it offered all in 1, and it is rarely the case when the system offers task management, an RMA form and drawing up contracts. The provider was flexible, with many implementations, adjusting the functionality to the needs of the company, however, it cost a lot

5   Discussion

The ability of a micro enterprise to recognize and understand the benefits of using ICT can make it increase the uptake and the use of ICT [8, 23]. Contrary to the results of the earlier research performed in Poland [13], the e-commerce micro enterprise feels the need of implementing the e-CRM system and it sees a potential in its application. It must come from the specific nature of the trade and the fact that its employees, normally, use the ICT and, according to Hung et al. [24], *"if employees in small organizations have more knowledge of information systems, then they will be more likely to adopt the information systems."*

The activity of an e-commerce micro enterprise focuses on providing efficient customer service, and that requires good organization and coordination of the work of all the employees participating in the process. One can assume that the reason is perceiving the e-CRM system as a tool aimed primarily at ensuring better work organization by using a module for managing the employee time and tasks, and not as a tool for customer service.

The results of the previous study by the Authors [25] show that the CEO does not appreciate a strategic approach to the CRM system implementation. At the same time the applicable literature recommends that micro enterprises should maximise the potential and minimise the risk of accepting the ICT by developing adequate strategies [28].

The study by Wynn et al. [44] performed in small enterprises has demonstrated that even in a small business enterprise, to be successful with the project, one must obey the principles of sound management. It shows that e-CRM implementations cannot allow a shortcut. Such implementations must be adapted to the general business strategy, they require a support of the top management, a clear vision of the project objectives, reasonable project management, a new way of thinking about the supply chain and adequately qualified and competent personnel. Unfortunately, the micro enterprise under study took a shortcut twice, assuming that it was enough to install software and to start using it. Even though, with the second attempt, the decision was taken to get subscription-charged software supplied by the software development company which was to provide training, more functionalities and a post-implementation support. However, there was no accurate pre-implementation analysis; the implementation conditions and all the costs of implementation were not identified, which is indicated as one of the critical success factors for IT implementations in SMEs [13].

The company's approach to implementing the e-CRM system confirms that small businesses tend to opt for a just-doing approach rather than for formal planning [26], also those in the e-commerce sector [11]. The integration of e-CRM with existing processes and structures was not prepared either, which is considered one of the conditions of an efficient use of the CRM solution [9].

The enterprise committed also other mistakes provided for in the literature as causes of failure [10, 29, 38]. The actual support from the top management was missing. The article co-author, who participated in the first attempt of implementation, pointed to the lack of such support, and the other interview with the CEO demonstrates that he knew little about the actions of the Digital manager, the person he made responsible for the system implementation. Another important, mentioned in the literature, cause for the failure is to think of CRM as pure technology. And it is exactly the approach of the

company to the CRM implementation which is revealed in the interviews. Both the CEO and the Digital manager mostly focused on software, its intuitiveness and functionality. Those are definitely important success factors, however they are insufficient.

On top of the mistakes made, there were also the traps the company did not expect, e.g., the provider supplying the demo the functionality of which differed from the full commercial software system version as well as the additional quite-high initial implementation costs and high customisation fees for additional functions or for integration with other systems used in the enterprise. A good communication between the implementation enterprise and the employees, considered one of the IT project success factors, was also missing [13].

The case has also confirmed that the e-CRM system availability in a computing cloud and provided in the SaaS model is very important and even of key importance for taking an implementation decision, which has been demonstrated in the previous studies in micro enterprises, small- and medium-sized enterprises [7, 31]. Such informatization model makes it possible to avoid the financial barriers and transfers the burden of software upgrade to the system provider.

6 Conclusions

The e-commerce enterprise has made two unsuccessful e-CRM system implementation attempts. The results of the study show that the first attempt did not provide any adequate conclusions which could facilitate the success of the second implementation attempt. Even though the second approach differed slightly from the first one, as the subscription-charged system and the provider which was to meet the company's expectations were chosen, however, again, a shortcut was taken by disregarding the important implementation success factors and by committing the mistakes provided for in the literature.

The company has also fallen in unexpected traps: a lack of information about the initial implementation costs (only a licence fee), a demo version incompliant with the full version, a hard-to-get-in-touch situation due to pandemic. However, in the opinion of the Authors, it shows that the enterprise was not prepared enough for the project as it did not identify the implementation conditions carefully enough.

The study, therefore, reveals that the company cannot learn from its own mistakes and, more importantly, the people do not understand what mistakes were made. Both the CEO and the Digital manager claim that the system is to be intuitive enough to be effectively used from the start. At the same time, they expect a wide functionality supporting quite complex customer service processes. An abundant functionality which is to facilitate those processes must result in a high complexity of the entire system. A question arises whether it is possible for the e-CRM software to offer an abundant functionality and to be, at the same time, intuitive enough for the training not to be needed. To answer the question, the Authors intend to perform yet another study with micro enterprises by looking for a case which would confirm it.

The study provided detailed information on the e-CRM software features desired by an e-commerce micro enterprise and offering a desired functionality. The solution is to operate in cloud computing offered in Polish, in the SaaS model, with a subscription charge and with low initial implementation costs. The system functionality should be possibly most intuitive and should include customer service, task allocation, calendar, customer complaints and servicing, payment processing, drawing up contracts and HR modules, with the mobile app, an internal communicator and a big cloud storage for files as a must.

Finally, let us be a little bit optimistic by saying that the enterprise did not get discouraged by two unsuccessful attempts and it took the third e-CRM system implementation attempt. The Digital manager still believes that, thanks to e-CRM, it will be possible to make the communication processes more efficient (cutting down on the number of emails) and to facilitate the task management as well as to have all the information needed in one easily-accessible place.

The Authors of the article are aware that the analysis of a single case is not sufficient enough to make some result generalizations, however the study can also provide a springboard for a bigger study in e-commerce micro and small enterprises which have made some e-CRM system implementation attempts.

References

1. Ahani, A., Rahim, N.Z.A., Nilashi, M.: Forecasting social CRM adoption in SMEs: a combined SEM-neural network method. Comput. Hum. Behav. **75**, 560–578 (2017)
2. Alford, P., Page, S.J.: Marketing technology for adoption by small business. Serv. Ind. J. **35**(11–12), 655–669 (2015)
3. Almotairi, M.: A framework for successful CRM implementation. In: Proceedings of the European and Mediterranean Conference on Information Systems, pp. 1–14 (2009)
4. Alpar, P., Reeves, S.: Predictors of MS/OR application in small business. Interfaces **20**(2), 2–11 (1990)
5. Ashley, C., Tuten, T.: Creative strategies in social media marketing: an exploratory study of branded social content and consumer engagement. Psychol. Market. **33**(1), 15–27 (2015)
6. Ashurst, C., Cragg, P., Herring, P.: The role of IT competences in gaining value from e-business: an SME case study. Int. Small Bus. J. **30**(6), 640–658 (2012)
7. Baker, O.: The adoption of cloud computing CRM in SME's, Southland, New Zealand. In: Proceedings of the IEEE Conference on Open Systems (ICOS), pp. 1–6 (2020)
8. Bharati, P., Chaudhury, A.: Studying the current status: examining the extent and nature of adoption of technologies by micro, small and medium sized manufacturing firms in the greater Boston area. Commun. ACM **49**(10), 88–93 (2020)
9. Boulding, W., Staelin, R., Ehret, M., Johnston, W.J.: A customer relationship management roadmap: what is known, potential pitfalls, and where to go. J. Market. **69**, 155–166 (2005)
10. Chalmeta, R.: Methodology for customer relationship management. J. Syst. Softw. **79**, 1015–1024 (2006)
11. Chaston, I., Badger, B., Mangles, T., Sadler-Smith, E.: Relationship marketing, knowledge management systems and e-commerce operations in small UK accountancy practices. J. Market. Manage. **19**(1/2), 109–129 (2003)
12. Chibelushi, C., Costello, P.: Challenges facing W. Midlands ICT-oriented SMEs. J. Small Bus. Enterprise Dev. **16**(2), 210–239 (2009)
13. Cieciora, M., Bołkunow, W., Pietrzak, P., Gago, P., Rzeźnik-Knotek, M.: Critical success factors of ERP/CRM implementation in SMEs in Poland: pilot study. Zeszyty Naukowe. Organizacja i Zarządzanie/Politechnika Śląska, (148 Contemporary management), pp. 103–116 (2020)

14. Dandridge, T., Levenburg, N.: High-tech potential? An exploratory study of very small firms' usage of the internet. Int. Small Bus. J. **18**(1), 81–91 (2000)
15. Daniel, E., Grimshaw, D.: An exploratory comparison of electronic commerce adoption in large and small enterprises. J. Inf. Technol. **17**(3), 133–147 (2002)
16. De Berranger, P., Tucker, D., Jones, L.: Internet diffusion in creative micro businesses: identifying change agent characteristics as critical success factors. J. Organ. Comput. Electr. Commer. **11**(3), 197–214 (2001)
17. Desouza, K.C., Awazu, Y.: Knowledge management at SMEs: five peculiarities. J. Knowl. Manage. **10**(1), 32–43 (2006)
18. eCommerce Innovation Centre: eCommerce in Welsh SMEs: The State of the Nation report 2005/6. Cardiff: eCommerce Innovation Centre, Cardiff University (2005)
19. Fillis, I., Wagner, B.: E-business development. Int. Small Bus. J. **23**(6), 603–634 (2005)
20. Garcia, I., Pacheco, C., Martinez, A.: Identifying critical success factors for adopting CRM in small: a framework for small and medium enterprises. In: Lee, R. (eds.) Software Engineering Research, Management and Applications, vol. 430, pp. 1–15. Springer, Heidelberg (2012). https://doi.org/10.1007/978-3-642-30460-6_1
21. Harrigan, P., Ramsey, E., Ibbotson, P.: Critical factors underpinning the e-CRM activities of SMEs. J. Market. Manage. **26**(13/14), 1–27 (2011)
22. Hill, J., Tiu Wright, L.: A qualitative research agenda for small to medium sized enterprises. Market. Intell. Plan. **19**(6), 432–443 (2001)
23. Hughes, M., Golden, W., Powell, P.: Inter-organisational ICT systems: the way to innovative practice for SME? J. Small Bus. Enterpr. Dev. **10**(3), 277–286 (2003)
24. Hung, S.Y., Hung, W.H., Tsai, C.A., Jiang, S.C.: Critical factors of hospital adoption on CRM system: organizational and information system perspectives. Decis. Support Syst. **48**(4), 592–603 (2010)
25. Januszewski, A., Krupcala, K.: Impact of the CRM system and time management on organizational effectiveness and performance: case study of an E-commerce micro enterprise. Eur. Res. Stud. J. **24**(1), 1157–1172 (2021)
26. Jones, C., Hecker, R., Holland, P.: Small firm Internet adoption: opportunities forgone, a journey not begun. J. Small Bus. Enterpr. Dev. **10**(3), 287–298 (2003)
27. Jones, P., Packham, G., Beynon-Davies, P., Pickernell, D.: False promises: E-business deployment in Wales' SME community. J. Syst. Inf. Technol. **13**(2), 163–178 (2011)
28. Jones, P., Simmons, G., Packham, G., Beynon-Davies, P., Pickernell, D.: An exploration of the attitudes and strategic responses of sole-proprietor micro-enterprises in adopting information and communication technology. Int. Small Bus. J. **32**(3), 285–306 (2014)
29. Kale, S.H.: CRM failure and the seven deadly sins. Market. Manage. **13**(5), 42–46 (2004)
30. Kennedy, A.: Electronic customer relationship management (eCRM): opportunities and challenges in a digital world. Irish Market. Rev. **18**(1–2), 58–69 (2006)
31. Khayer, A., Talukder, M.S., Bao, Y., Hossain, M.N.: Cloud computing adoption and its impact on SMEs' performance for cloud supported operations: a dual-stage analytical approach. Technol. Soc. **60**, 1–15 (2020)
32. Khlif, H., Jallouli, R.: The success factors of CRM systems: an explanatory analysis. J. Global Bus. Technol. **10**(2), 25–42 (2014)
33. Lecerf, M., Omrani, N.: SME internationalization: the impact of information technology and innovation. J. Knowl. Econ. **11**(2), 805–824 (2020)
34. Lees, J.D.: Successful development of small business information systems. J. Syst. Manage. **38**(9), 32–39 (1987)
35. Levy, M., Powell, P.: Emerging technologies: can the Internet add value for SMEs. In: Proceedings of the Fourth United Kingdom Association of Information Systems Conference, University of York (1999)

36. Lind, M.R., Zmud, R.W., Fischer, W.A.: Microcomputer adoption—the impact of organizational size and structure. Inf. Manage. **16**(3), 157–162 (1989)
37. Matarazzo, M., Penco, L., Profumo, G., Quaglia, R.: Digital transformation and customer value creation in Made in Italy SMEs: a dynamic capabilities perspective. J. Bus. Res. **123**, 642–656 (2021)
38. Nguyen, T.H., Sherif, J.S., Newby, M.: Strategies for successful CRM implementation. Inf. Manage. Comput. Secur. **15**(2), 102–115 (2007)
39. Owens, I., Robertson, D.: Aligning e-commerce with business strategy: the case of the Bank of Scotland. In: Proceedings of the fifth United Kingdom Association of Information Systems Conference, University of Wales Institute, Cardiff, 26–28 April (2000)
40. Padilla-Meléndez, A., Garrido-Moreno, A.: Customer relationship management in hotels: examining critical success factors. Curr. Issues Tour. **17**, 387–396 (2014)
41. Simmons, G., Armstrong, G., Durkin., M.: An exploration of small business website optimization: enablers, influencers and an assessment approach. Int. Small Bus. J. **29**(5), 534–561 (2011)
42. Welsh, J.A., White, J.F.: A small business is not a little big business. Harv. Bus. Rev. **59**(4), 18–32 (1981)
43. Wolcott, P., Kamal, M., Qureshi, S.: Meeting the challenges of ICT adoption by micro-enterprises. J. Enterp. Inf. Manage. **21**(6), 616–632 (2008)
44. Wynn, M., Turner, P., Banik, A., Duckworth, G.: The impact of customer relationship management systems in small business enterprises. Strat. Change **25**(6), 659–674 (2016)
45. Yin, R.K.: Case Study Research: Design and Methods. Applied Social Research Methods Series, vol. 5. Sage Publications (2009)
46. Ziff Davies Research: Leveraging CRM for MidSize Company Growth. White Paper. Ziff Davies Inc. and Intelligent Business Strategies, New York (2012)

The Diagnostic Model for Assessing the State of Stability of an Industrial Enterprise

Olena Rayevnyeva[1] (iD), Mikolaj Karpinski[2](✉) (iD), Olha Brovko[1] (iD), Pawel Falat[2] (iD),
and Iryna Aksonova[1] (iD)

[1] Simon Kuznets Kharkiv National University of Economics, Kharkiv, Ukraine
[2] University of Bielsko-Biala, Bielsko-Biala, Poland
{mkarpinski,falat}@ath.bielsko.pl

Abstract. At the present stage, Ukraine is in a state of market transformation which is accompanied by the spread of global integration processes, digitalization of social and economic processes, open national economies and, consequently, the intensification of crises. In these conditions, the formation of effective tools for diagnosing the state of industrial enterprises will identify timely latent threats to its sustainable behaviour, overcome the crisis if it occurs, and ensure its revival at the same or higher level of organization and efficiency. The violation of cyclicality (failure to overcome the crisis) leads to the termination of its activities as a business entity. The article presents an algorithmic model for diagnosing the class of resilience/crises of industrial enterprises and develops the diagnostic scales that serve as a tool for recognizing the current and future state of the enterprise. It is proposed to study the activity of the enterprise as a set of its system-forming spheres - production, financial and labour. The diagnostic scales are developed on the basis of processing of the information on the allocated spheres for 24 enterprises of the industry of Ukraine for 10 years with application of information and computer technologies. They serve as a basis for the formation of adequate management impulses and the construction of achievable scenarios for crisis management in the enterprise, taking into account its resource-capabilities. The methods of correlation, taxonomic and cluster analyzes are mathematical tools for constructing scales. The proposed scales are universal and can be used as an analogue of industry standards.

Keywords: Crisis situation · Enterprise · Diagnostic model · Scale · Sphere of enterprise's life

1 Introduction

The development of the modern economy of Ukraine is characterized by a rapid pace of economic and technological transformations, which is intensified by the competition in its industries in accordance with the dynamic changes in consumer needs. However, the domestic business sphere shows the signs of prolonged stagnation, to overcome which is a critical task for the development of the Ukrainian economy at both the macro and micro levels.

© Springer Nature Switzerland AG 2021
S. Wrycza and J. Maślankowski (Eds.): PLAIS EuroSymposium 2021, LNBIP 429, pp. 51–67, 2021.
https://doi.org/10.1007/978-3-030-85893-3_4

According to the liberal concept of economics, there are special conditions for business entities. However, it should be noted that industrial enterprises are one of the subjects of the economy, which is strongly influenced by economic cycles and competition in the markets. This is especially true of business development in the context of digital transformation, which is a new dimension of reality with the definition of digital leadership.

In today's world, digital technologies create fundamentally new opportunities for building interaction between government, business and the public, eliminating long chains of intermediaries and accelerating the conduct of various transactions and operations. Such factors come to the fore due to the rapid development of information technology and globalization of the economy. They offer fundamentally new concepts of consumption and open up additional potential for the development of new markets and access to them. In such conditions, organizations need to intensify their development to be innovative, so as not to lose competitiveness and communication with consumers.

The solution of this problem belongs to the type of complex multicriteria tasks, which actualizes scientific and economic research in diagnosing the trajectory of enterprise development, its resource capabilities, business management system. These issues become especially relevant in conditions of force majeure, particularly in the context of the COVID-19 pandemic. Due to the crisis caused by the COVID-19 pandemic, Ukrainian business in 2020 is on the verge of survival. The economic downturn, declining purchasing power and changing behaviour of citizens, weak support (or inefficiency) by the state, have led to the reduction in production and the creation of social tensions in society. All this necessitates the constant search for effective approaches, tools, tools for monitoring and diagnosing the state of business.

The article's purpose is to develop a diagnostic model for assessing the state of sustainability of an industrial enterprise in the crusial spheres of its life. This model is aimed at recognizing trends in the development of an enterprise in accordance with the changes in the external and internal environment.

To achive the goal of the article, the following tasks are solved:

1) Formation of the information space of recognition of a class of stability/crisis at the enterprise.
2) Development of a criteria for recognizing enterprises that are in a state of resilience/crisis
3) Forming the rule of recognition of stable/crisis enterprises
4) Grouping of industrial enterprises of Kharkiv region.
5) Developing the ranges of values of indicative indicators of recognition of a class of stability/crisis at the enterprise.

2 Literature Review

Today, there is a large number of scientific papers, which at the theoretical, methodological and practical levels explored a wide range of issues related to the diagnosis of enterprises. Diagnosis is often associated with diagnosing only the financial condition of the company.

V. Ponomarenko, O.Tridid (Ponomarenko 2002), E. Utkin (Utkin 1997) and others made a significant contribution to the study of enterprise economics in crisis conditions and to the theory of crisis management. In the works of these authors, considerable attention is paid to the efficiency and safety of the enterprise, the development of theory and practice of diagnosing the crisis and the threat of bankruptcy.A significant number of factors, which affect the financial sustainability of the enterprise, cause the lack of a single approach to its definition and assessment in the economic literature (Yalovy and Bakerenko 2011).

Khrystynko and Butkova (2011) considered sustainable economic development, mainly at the level of the country or region. One of the elements of economic sustainability of the enterprise is the personnel sustainability. The value of this is that the level of qualification and competence of employees determines the enterprise competitiveness in the labour market. Therefore, it determines its ability to attract staff on time, to form a team with the necessary characteristics for the enterprise, to update the staff.

Galyna Azarenkova, Olena Golovko and Kateryna Abrosimova (2018) believed that a financial sustainability of the enterprise is a key feature of its financial status, its strategic development. Timely analysis of financial sustainability creates new opportunities for the enterprise to identify reserves in order to enhance its competitive position, increase market share and fulfill other tactical and strategic goals.

Monetary and financial stability of the enterprise in the coronavirus outbreak became Tobias' research subject (2020). The assumption was made that if economic and financial conditions were to deteriorate further, policymakers could revert to the broader toolkit developed during the financial crisis.

Iryna Trunina, Denys Zagirniak and other authors (2020) study the sustainability in different contexts. Some identify the indicators and predictors of the sustainability, and others identify the need to ensure personnel sustainability as a competitive advantage of the enterprise.

The traditional methods of financial sustainability assessment can be divided into three groups. The first group includes the qualitative assessment of financial stability, the second one contains the quantitative assessment of financial stability, and the third group includes the assessment of financial insolvency of enterprises. Among the methods of enterprise's financial state analysis and its' financial sustainability the following methods can be outlined: time series models, regression models, models of the systems of interrelated variables, recursive systems, etc. (Zakharova 2013).

In the work of V.Savchuk (Savchuk 2020) it is presented a set of issues of financial diagnostics and monitoring of the enterprise from the standpoint of making sound management decisions. The issues of financial analysis are considered in their organic relationship with strategy and marketing, as well as with internal business processes of the enterprise.

A number of researches are devoted to the development of conceptual and methodical bases of diagnostics of various kinds of activity of the enterprise separately. For example, in Ruslan Skrynkovskyy and Oksana Klyuvak's work (2016), the authors pay attention to the diagnosis of the export potential of the enterprise. They determine that to analyze and assess the level of potential and readiness of the enterprise to export activities, it is necessary to form a system of business indicators. These combine indicators of overall competitiveness in international and national markets, the level of product competitiveness based on its resource capabilities.

Bilal Bin Saeed and Wenbin Wang (2013) focus on the organizational diagnostics. Based on a critical analysis of existing models, the authors proposed a new model of organizational diagnostics based on three criteria - the model should be clear and not very complex, it should meet the specifics of the organization, the diagnostic model should collect data in the diagnostic process.

The research of Kumar, Maneesh and Harris, Irina (2020) is devoted to the improvement of diagnostic methodology in the enterprise as a whole. It is proved that total diagnostics increases the efficiency of integration between the levels of organization and departments, and requires the development of information model basis.

The issues of improving the methods of analysis and diagnosis in the process of enterprise management are also relevant in the research of scientists. So, O. Rayevnyeva and other authors (2020) think that a diagnostic is an effective tool for managing enterprise behaviour taking into account the stage of its business cycle. The resource capabilities of the enterprise correspond to the stages of the cycle, the cognitive relationship of the main system-forming areas of the enterprise. By analyzing and evaluating these capabilities, the opportunity of crisis management is determined - the possibility of provoking occurrence of artificial, positive crises or elimination of the natural, negative crises.

Zachosova and Babina (2018) simulate the behaviour of Ukrainian financial institutions' economic security in 2018 on the conjuncture of the financial market and the state financial security.

The influence of industry on the economic development of Ukraine is studied by the authors N. Marynenko (2016), M. Sushko (2017), O. Bilska (2020), V. Gurochkina (2020). Their works reveal the reasons for the low level of profitability, business activity, insolvency of industrial enterprises, the wave-like nature of the dynamics of profitability.

L. Deineko, V. Zymovets, N. Sheludko and others (2018) argue that industry is the main driver of economic growth of the Ukrainian economy. Thus, statistical studies of the industrial production index and the gross domestic product index show a one-way trend of change in these indicators. They also prove the conclusion of a significant contribution of the Ukrainian industrial enterprises to general economic development.

However, the studies have shown that the scientific literature reveals only some aspects of the diagnosis, and often diagnosis is defined as a direction of economic analysis which significantly narrows the focus and scope of its action.

Developing the existing approaches to diagnostics of functioning of the enterprise, diagnostic scales of a condition of development of the industrial enterprise are offered in work. They are based on the basic system-forming spheres of its vital activity - industrial, financial and labour. This allows to determine the current and future trajectory of the enterprise, as well as to form a general direction of managerial influences to adjust the future behaviour of the enterprise taking into account its resource capabilities. This position of the authors is justified by the following considerations. The activity of an industrial enterprise is a set of various spheres of its life activity, starting from a purely production and ending with marketing and sales activities and the formation of the image of the enterprise. All these areas are important for a stable enterprise in its synergetic combination. But if the company operates for a long period of time in the conditions of severe financial constraints caused by constant political and economic destruction, the key areas of its activities are production, labour and financial spheres. They allow all other spheres of life to function effectively.

3 Problem Description and Basic Assumptions

This section introduces the research problem in this study and presents the assumptions of the scientific research.

3.1 Problem Description

Modern complex open socio-economic systems, enterprises are not isolated. They operate under the conditions of active influence of the external environment and are forced to adapt accordingly to its changes based on their resource capabilities. In this regard, the diagnosis of the main trends of changes in the functioning of enterprises in market conditions is becoming more relevant, which is strategically important for both specific regions and for the whole country. There is an urgent need to manage crisis situations of industrial enterprises through the formation of balanced, adequate to these changes trajectories of the further development of enterprises.

Over the past few years, most businesses have been in crisis (Fig. 1).

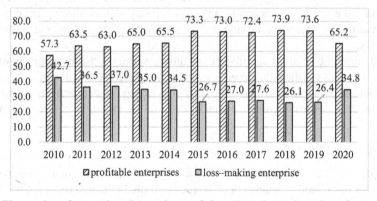

Fig. 1. The number of enterprises that made a profit/loss (% to the total number of enterprises/% to the total) (Source: State Statistics of Ukraine)

Presented in Fig. 1 data show that the number of unprofitable enterprises in Ukraine is gradually increasing every year. The graph shows that there is a tendency to increase the share of unprofitable enterprises in the period from 2018 to 2020. In 2020, compared to 2019, the number of enterprises that suffered a loss increased by 1.32%. This is due to the introduction of quarantine restrictions, slowing down production volumes, reduction of the demand for products/services.

Since industrial enterprises are subjects of the economy and are strongly exposed to the economic cycles and competition in the markets, so these enterprises are sensitive to

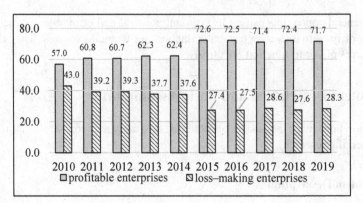

Fig. 2. The number of industrial enterprises that made a profit/loss (% to the total) (Source: State Statistics of Ukraine)

crises. Figure 2 shows the dynamics of profit/loss of industrial enterprises for the period 2010–2019.

Such crisis phenomena have not only a negative economic but also a social effect, which provokes increasing tensions in society. Thus, during the period 2010–2020, the number of employees in business entities decreased by 17%.

In this regard, the diagnosis of major trends in industrial development in an unstable environment is one of the urgent tasks of the enterprise management. It requires a constant search for the effective forms and means of processing and monitoring large amounts of information by means of economic and mathematical modeling and information technology.

In addition, the result of an effective diagnostic system is the development of management solutions to adjust the trajectory of the enterprise, localization of crises, taking into account its resources and bankruptcy. This is a strong signal to actively interact with its stakeholders - owners, workers, financial institutions, investors, contractors.

3.2 Basic Assumptions

Assumption 1. The stability of the behaviour of enterprises in the market environment depends on the effective approaches, diagnostic tools of its enterprise, based on the processing of large arrays of information by means of information and communication technologies.

Assumption 2. The state of the enterprise is a set of spheres of its system-forming spheres of life, namely financial, labour, production, which interact with each other and determine its market development in accordance with its resource capabilities.

Assumption 3. The quality of the diagnostics system of the internal and external environment of the enterprise should be based on the use of adequate economic and mathematical methods, approximation of trends in system-forming spheres of its life and serve as a basis for effective management decisions to adjust market behaviour.

4 Model Formulation

This section presents an algorithmic model for diagnosing the behaviour of industrial enterprises and a description of a set of economics-mathematical methods and models that act as diagnostic tools.

4.1 Notation

Diagnosis is an integral part of the management system of each enterprise, as it aims to identify retrospective, current and future status, as well as the development of preventive, remedial and reactive management solutions. These are aimed at eliminating problems and using the chances of the operating environment.

The proposed algorithm for recognizing the class of stability/crisis of the industrial enterprises and the formation of indicative values of indicators that characterize this phenomenon is presented in Fig. 3.

To build diagnostic scales for recognizing the types of enterprises (Fig. 3, stage 5), the use of the following classes is proposed:

1. For the type of sustainable enterprises
   the class of absolute stability of the enterprise, when stocks and costs are less than the amount of working capital and bank loans for inventory. Indicators that characterize the provision of production assets and labour productivity, show a stable tendency to increase. The structure and number of production staff is stable. The share of highly skilled workers and average wages are growing;
   the class of normal stability, when the solvency of the enterprise is guaranteed, capital adequacy and capital efficiency are at a high level, the share of material costs in the cost of production increases, and profitability tends to decrease. There is a stable share of employees with higher and secondary special education, increase in the minimum wage in the country;
   the class of unstable (pre-crisis) state, when there is a violation of solvency, but the possibility of reproducing the balance of means of payment and payment obligations by attracting temporarily free sources of funds in the turnover of the enterprise. There is a slowdown in fixed assets growth and a predominant increase in material costs, the growth rate of the average wage slows down. The turnover ratio for hiring employees decreases and the turnover ratio for retirement increases.
2. For crisis enterprises
   the class of slight crisis, when there are significant stocks of finished products in the warehouse. There is reducing turnover from sales, rising costs, slowing down the dynamics of fixed assets and their turnover. The share of material costs in total continues to increase, the staff structure changes to increase the share of less skilled workers. There is a tendency to reduce productivity;
   the middle crisis class, when there is the absence or insignificant level of insurance (reserve) funds, a high level of accounts payable, large amounts of low-liquid current assets and a large amount of investment with a long payback period. The capital adequacy and return on assets is low, the return on fixed assets is close to zero. The

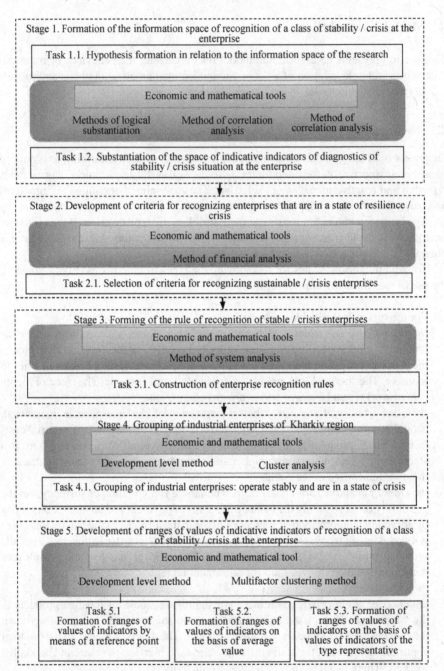

Fig. 3. The algorithm for diagnosing the state of development of an industrial enterprise (Source: made by the authors)

average wage is constantly decreasing, the turnover ratio of retirement of employees is significantly higher than the turnover ratio of the reception of employees;
the class of severe crisis, when the share of equity is less than the share of borrowed capital. There is a lack of equity for further investment in production, material consumption exceeds one, and the return on fixed assets may become negative. There is a significant decline in productivity and average wages, and debt to staff accumulates.

In order to eliminate the zones of uncertainty in the scales between the criteria values of indicators of certain classes, the following procedure is proposed:

1. Determining the interval of uncertainty between adjacent classes by the formula:

$$N = X^B_{min} - X^H_{max} \tag{1}$$

where X^B_{min} - the lower value of the indicator of the highest class of the scale; X^H_{max} - the upper value of the indicator of the lower class of the scale.

2. Determining the average value of the uncertainty zone:

$$\bar{x} = \frac{N}{2} \tag{2}$$

3. Formation of criterion values of indicative indicators for middle classes:

   a) for indicators of stimulants:

$$X^H_B = X^B_{min} - \bar{X}; \quad X^H_B = X^H_{max} + \bar{X} \tag{3}$$

   b) for indicators of destimulators:

$$X^B_H = X^B_{min} + \bar{X}; \quad X^H_B = X^H_{max} - \bar{X} \tag{4}$$

where X^B_H- the lower limit of the highest class of the scale;
X^H_B- the upper limit of the lower class of the scale.

5 Solution Method

Within the limits of this section the complex of economic and mathematical tools, which is used for construction of diagnostic scales of recognition of a class of stability/crisis of the industrial enterprise, is presented.

5.1 Basis for Method Selection

Let's consider the economic and mathematical tools used to solve each problem.

Task 1.2. Substantiation of the space of indicative indicators of diagnostics of stability/crisis situation at the enterprise. In solving this problem such methods were used:

a) the method of logical substantiation, monographic, comparative, content analysis for the formation of the primary list of indicators of the indicative space of the study of the industrial enterprise as a whole and its system-forming spheres of life;

b) the method of correlation analysis to eliminate duplication of information on the basis of determining the correlations between the indicators of the primary indicative space of the study of the internal environment of the enterprise as a whole and by areas of life. The pairwise correlation coefficient was used for this:

$$r = \frac{\sum\limits_{i=1}^{n}(x_i - \bar{x})(y_i - \bar{y})}{\sqrt{\sum\limits_{i=1}^{n}(x_i - \bar{x})^2}\sqrt{\sum\limits_{i=1}^{n}(y_i - \bar{y})^2}} \tag{5}$$

where x_i, y_i – the value of the levels of factor and performance indicators;
$\bar{x}$, $\bar{y}$ – average values of the levels of indicators.

Task 2.1. Selection of criteria for recognizing sustainable/crisis enterprises. Mathematical tools for solving the problem are the methods of financial analysis. They allow you to assess the current and future financial condition of the enterprise; determine the possible appropriate pace of the development of enterprise from the standpoint of financial security; identify available sources of funds and assess the possibility and feasibility of their mobilization, forecast the company's position in the capital market.

The primary external manifestation of the crisis is the formation of a steady trend of increasing current costs, as well as reducing the volume of activity, income and profits. Further deepening of the crisis is characterized by a catastrophic deterioration of all the indicators of its condition (both quantitative and qualitative), which leads to a gradual loss of equity (net assets) and a shortage of financial resources to settle liabilities. Based on the following judgments, it is proposed to choose the criteria of net profit/loss (CPR) and the ratio of own funds (Kz.v.z) as a criteria for recognizing the crisis.

Task 3.1. Construction of the enterprise recognition rules. The solution of the problem is aimed at forming a motorcade of the above indicators of financial activity of the enterprise, namely:

$$S_{enterprise} = <CPR; Kz.v.z.>, \tag{6}$$

where CPR $= 1$, if there is a tendency to increase net income; CPR $= 0$, if there is a tendency to reduce net income;
Kz.v.z $= 1$, if the value of the coefficient is observed $\geq 0{,}1$; Kz.v.z $= 0$, if the value of the coefficient is observed $< 0{,}1$.

The rules for recognizing the level of stability of the enterprise are as follows:

1. $S_{enterprise} = \{1; 1\}$ provided that the state is maintained for at least 2/3 of the time of the analyzed period - the company is stable.
2. $S_{enterprise} = \{1; 1\}$ 2/3 of the time of the analyzed period - the company is stable.
3. $S_{enterprise} = \{0; 1\}$ provided that the state is maintained for at least 1/3 of the time of the analyzed period - a crisis enterprise.
4. $S_{enterprise} = \{1; 0\}$ provided that the state is maintained for less than 1/3 of the time of the analyzed period - a crisis enterprise.

To solve task 4.1, the taxonomic method of the level of development and cluster analysis are used, the combination of which allows to form 2 clusters of enterprises: enterprises that function stably in a market economy and enterprises in crisis. The integrated indicator of the stability of the enterprise is proposed to be calculated on the basis of the method of the level of development - the method of taxonomic analysis developed by Z. Helwig (Pluta 1980). The choice of this method is justified by the fact that the integrated indicator accumulates the influence of various indicators in the financial, labour and production spheres. It allows to give an economic description of the behaviour of the enterprise from a systemic standpoint. The values of the indicator are normalized and range from 0 to 1. The economic interpretation of the values of the integrated indicator is as follows: the closer the value is to 1, the more stable the company operates.

The calculation of the integrated indicator (Di) is as follows:

$$D_i = 1 - \frac{C_{i0}}{C_0} \tag{7}$$

$$C_0 = \overline{C}_0 + 2 * S_0; \qquad S_0 = \sqrt{\frac{\sum_{i=1}^{w} (C_{i0} - \overline{C}_0)^2}{w}}; \qquad \overline{C}_0 = \frac{\sum_{i=1}^{w} C_{i0}}{w}$$

where D_i – an indicator of enterprise/sphere of life development;

To solve tasks 4.2 and 5.2–5.3, the paper proposes the use of cluster analysis, namely the method of k-means. This method should be used when there are certain a-priori hypotheses about the number of clusters. The advantage of this method is to obtain disparate cluster groups with unambiguous economic interpretation.

6 Computational Case

The approbation of the proposed algorithmic model was carried out on the example of 24 industrial enterprises of Ukraine for 10 years.

The result of stage 1 is an economically sound system of indicators that reflects the development of the enterprise as a whole and in the spheres of its life (Fig. 4).

The obtained system of indicators is the basis for the development of diagnostic scales for recognizing the class of resilience/crisis of an industrial enterprise.

As a result of solving the tasks of stage 2 (Fig. 2), the indicators of net profit/loss and the ratio of own funds were selected as criteria for recognizing the crisis. The choice of these criteria is justified by the fact that:

1) the amount of net profit/loss provides users, primarily investors and creditors, important financial information to assess the past performance of the enterprise, as well as the risk of not achieving the expected results;

2) the ratio of own funds is considered as a criterion for determining the insolvency (bankruptcy) of the enterprise and characterizes the availability of working capital of the enterprise, necessary for its financial stability.

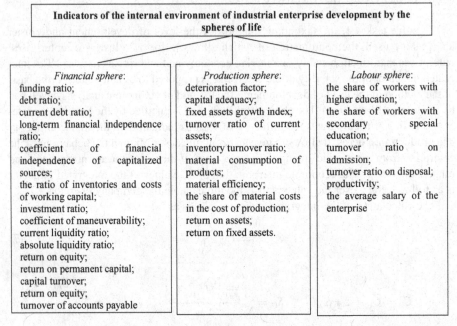

Fig. 4. Indicators of industrial enterprise development by the spheres of life

By using the proposed rules for the recognition of industrial enterprises (Fig. 3, stage 3), the division of 24 industrial enterprises of Ukraine for 10 years on an annual basis was conducted. The analysis showed that half of the enterprises are in a state of crisis, and the others are stable (Table 1).

The result of solving task 4.1 is the construction of an indicator of the enterprise development in the spheres of its life in Table 2.

The obtained values of the integrated indicator for each sphere of life of the enterprise served as an information basis for the construction of diagnostic scales for recognizing the class of stability/crisis of the industrial enterprise.

The result of stage 4 (Fig. 3) is the grouping of industrial enterprises that are in a state of stability into 3 clusters. Out of 120 situations of functioning of the industrial enterprises, 38 situations are referred to the first cluster; 70 situations - to the second cluster; 12 situations - to the third class.

Table 1. The result of grouping industrial enterprises of Ukraine

№	Sustainable industrial enterprises	Industrial enterprises in crisis
1	OJSC "Experimental Electrical Plant"	OJSC Electrotechnical Plant "Ukrelektromash"
2	OJSC Kharkiv Machine-Building Plant "Light of Shakhtar"	OJSC "Electromashina"
3	State Enterprise Electrovazhmash Plant	OJSC "Kharkiv Plant of Aggregate Machines"
4	OJSC "Kharkiv Plant of Aggregate Machines"	OJSC "Kharkiv Plant of Aggregate Machines"
5	OJSC "Frunze Plant"	OJSC "Avtramat"
6	CJSC Lozovsky plant "Traktorodetal"	OJSC "Izium Locomotive Repair Plant"
7	CJSC Kharkiv Order "Badge of Honor" Machine-Building Plant "Red October"	CJSC "Kharkiv plant of electrical products №1"
8	OJSC Kharkiv plant "Tochmedprilad"	OJSC "Ordzhonikidze Kharkiv Tractor Plant"
9	OJSC Kharkiv Electrotechnical Plant "Transvyaz"	OJSC Kharkiv Bearing Plant
10	OJSC Kharkiv Electric Equipment Plant	CJSC Joint Venture "HEMZ-IRES"
11	CJSC "ELOX"	CJSC "Interconditioner"
12	PJSC "Finprofil"	DNVP "System"

According to the results of multifactor clustering, out of 120 situations of functioning of the enterprises that are in a state of crisis, 6 situations are referred to the first class; 18 situations - to the second class; 96 situations - to the third class.

The construction of diagnostic scales for recognizing the class of stability/crisis of an industrial enterprise in the financial, industrial and labour spheres is the result of solving the 5th stage (Fig. 3). Table 3 shows a fragment of diagnostic scales (criterion values of indicators) for recognizing the level of stability/crisis of the enterprise in the labour sphere.

Similar scales that include a complex of indicators have also been developed for the production and financial spheres of life of the industrial enterprises.

The obtained scales are an effective tool for recognizing latent and obvious problems in each area of life of the enterprise and the development of managerial influences for their elimination or localization.

The expediency of their practical application lies in the possibility of adjusting the trajectory of the enterprise as a whole on the basis of establishing specific values of the proposed indicators for the production, financial and labour spheres as planned quantitative benchmarks in accordance with the real resource capabilities of the enterprise.

Table 2. The integral indicator of the development of sustainable industrial enterprises by spheres of life (fragment)

Period	Integral in the financial sector	Integral in the manufacturing sector	Integral in the labor sphere	Period	Integral in the financial sector	Integral in the manufacturing sector	Integral in the labor sphere
1	0.116	0.169	0.117	10	0.222	0.210	0.364
	0.202	0.171	0.215		0.232	0.092	0.285
	0.131	0.174	0.201		0.137	0.176	0.352
	0.216	0.223	0.314		0.194	0.193	0.349
	0.189	0.213	0.292		0.194	0.268	0.314
	0.225	0.204	0.216		0.170	0.173	0.202
	0.168	0.182	0.209		0.229	0.177	0.314
	0.186	0.187	0.233		0.163	0.281	0.253
	0.138	0.174	0.194		0.170	0.193	0.104
	0.147	0.156	0.192		0.234	0.207	0.286
	0.268	0.262	0.285		0.243	0.293	0.361
	0.213	0.152	0.204		0.178	0.190	0.247

Table 3. The diagnostic scale for recognizing the class of resilience/crisis for the labour sphere of industrial enterprises

Stability/crisis class of the enterprise	The name of the indicator					
	HE (share of workers)	SSE (share of workers)	TOA	TOD	Prod	AS
Sustainable industrial enterprises						
absolute stability	0,562–0,380	0,225–0,315	0,296–0,231	0,198–0,249	4368,474–2038,666	4,137–3,210
normal stability	0,379–0,333	0,316–0,412	0,230–0,183	0,250–0,280	2038,665–738,666	3,209–1,896
unstable (pre-crisis)	0,332–0,261	0,413–0,444	0,182–0,150	0,281–0,350	738,665–263,782	1,895–1,174
Crisis industrial enterprises						
A slight crisis	0,447–0,352	0,190–0,239	0,588–0,280	0,206–0,438	2530,937–904,373	2,926–1,575
Medium crisis	0,351–0,272	0,240–0,282	0,279–0,220	0,439–0,648	904,372–425,873	1,574–1,249
Severe crisis	0,271–0,203	0,281–0,294	0,219–0,127	0,649–0,816	425,872–207,139	1,248–0,829

7 Conclusion

A characteristic feature of the current stage of development of the Ukrainian economy is, on the one hand, the recognition of the key role of industry, and, on the other hand, the existence of a deep systemic crisis of industrial enterprises. Therefore, one of the urgent tasks of the management of industrial enterprises is to create an effective and efficient system for recognizing the state of their stability. All this necessitates the improvement of existing tools, diagnostic tools.

Based on the studies of the functioning of 24 industrial enterprises of Ukraine, an algorithm for determining the general indicative values of financial, production and labour indicators of enterprises operating in conditions of stability and different depths of crisis phenomena, has been developed. Under the conditions that the developed ranges of value change of these indicators are defined on the basis of the analysis of the enterprises of mechanical engineering, they can be considered as the most characteristic values reflecting a certain class of stability/crisis at the enterprise. These values form the basis of the system of monitoring the activities of industrial enterprises. They can also be used as an information basis for the formation of effective management influences to adjust the trajectory of future behaviour of the enterprise. The future trajectory can be formed by provoking an artificial crisis to strengthen the upward trajectory of the enterprise development or natural crisis management in order to localize negative phenomena and create the preconditions for changing the general vector of the development from descending and ascending.

On the basis of the conducted researches, the algorithm of definition the general indicative values of the financial, industrial and labour indicators of the enterprises operating in the conditions of stability and various depth of display of the crisis phenomena is developed. Under the conditions that the developed ranges of change of the value of these indicators are defined on the basis of the analysis of the enterprises of mechanical engineering, they can be considered as the most characteristic values reflecting a certain class of stability/crisis at the enterprise. These values form the basis of the monitoring system of industrial enterprises. They can also be used as an information basis for the formation of the effective management influences to adjust the future behaviour of the enterprise, namely the formation of an artificial crisis to strengthen the upward trajectory of the enterprise development or natural crisis management changes in the general vector of development from descending and ascending.

The use of modern economic and mathematical tools for analyzing the behaviour of the enterprise as a whole and the main system-forming areas of its life significantly improves the quality of management decisions. These decisions are aimed at maintaining an upward or changing downward trajectory of the development by ascending a set of appropriate management responses to artificial or natural crisis. To this end, the study substantiates the use of the taxonomic method of the development level, cluster, correlation, financial and system analysis.

Further researches on the diagnosis of market behaviour of the industrial enterprises should be aimed at developing economic and mathematical tools for identifying promising trends in system-forming areas of the enterprise, creating cognitive models of interaction of certain indicators, developing a system to support management decisions to adapt to fluctuating the market environment.

References

Сушко, М.Ю.: Аналіз сучасного стану розвитку промислової галузі України. Вісник економічної науки України **1**(32), 93–98 (2017). http://dspace.nbuv.gov.ua/handle/123456 789/123076

Більська, О.В.: Аналіз промислового комплексу України. Ефективна економіка, № 8 (2020). http://www.economy.nayka.com.ua/?op=1&z=8125. https://doi.org/10.32702/2307-2105-2020.8.7.дата звернення: 24.06.2021

Гурочкіна, В.В.: Економічна динаміка розвитку промислових підприємств в економіці емерджентного типу. Вчені записки ТНУ імені В. І. Вернадського. Серія: Економіка і управління, Том. 31(70), №. 3, pp. 127–134 (2020). http://www.econ.vernadskyjournals.in.ua/journals/2020/31_70_3/31_70_3_1/23.pdf

Мартиненко, Н.Ю.: Тенденції розвитку промислових підприємств України в сучасних умовах. Ефективна економіка, № 1 (2016). http://www.economy.nayka.com.ua/?op=1&z=4724

Плюта, В.: Сравнительный многомерный анализ в экономических исследованиях: Методы таксономии и факторного анализа. Пер. с пол. В. В. Иванова; науч. ред. – В. М. Жуковской. – М: Статистика, 151 p (1980)

Пономаренко, В.С.: Стратегічне управління розвитком підприємства: навчальний посібник. Понаморенко, В.С., Пушкар, О.І., Тридід,О.М. Харків: ХДЕУ, 640 p (2002)

наук Дейнеко, Л.В.: Розвиток промисловості для забезпечення зростання та оновлення української економіки: науково-аналітична доповідь. за ред. д-ра екон. НАН України, ДУ «Ін-т екон. та прогнозув. НАН України». – К., 158 p (2018). http://ief.org.ua/docs/sr/301.pdf

Савчук, В.П.: Диагностика предприятия: поддержка управленческих решений [Электронный ресурс]. 3-е изд.(эл.)М.: Лаборатория знаний, 176 p (2020)

Уткин, Э.А.: Антикризисное управление. М.: Ассоциация авторов и издателей «Тандем», изд-во «Экмос», 400 p (1997)

Saeed, B.B., Wang, W.: Organisational diagnoses: a survey of the literature and proposition of a new diagnostic model. Inf. Syst. Change Manag. **6**(3), 222–238 (2013)

Azarenkova, G., Golovko, O., Abrosimova, K.: Management of enterprise's financial sustainability and improvement of its methods. Account. Financ. Control **2**(1), 1–14 (2018). https://doi.org/10.21511/afc.02(1).2018.01.   https://www.researchgate.net/publication/259497205_Organisational_diagnoses_A_survey_of_the_literature_and_proposition_of_a_new_diagnostic_model. Accessed 12 May 2021

Trunina, I., Zagirniak, D., Pryakhina, K., Bezugla, T.: Diagnostics of the enterprise personnel sustainability. Probl. Perspect. Manag. **18**(2), 382–395 (2020). https://doi.org/10.21511/ppm.18(2).2020.31

Khrystynko, L., Butkova, N.: Model ekonomichnoi stiikosti promyslovoho pidpriemstva [Model of economic stability of an industrial enterprise]. Bull. Khmelnytsky Natl. Univ. **5**, 131–134 (2011). (In Ukrainian)

Kumar, M., Harris, I.: Enterprise-wide diagnostic in the Uk SME: focus beyond tools and techniques. Prod. Plann. Control **32**(9), 730–746 (2020)

Rayevnyeva, O., Brovko, O., Filip, S., Aksonova, I., Derykhovska, V.: Management and modelling of the industrial enterprise's crisis situations. Probl. Perspect. Manag. **18**(1), 192–205 (2020). https://doi.org/10.21511/ppm.18(1).2020.17

Skrynkovskyy Ruslan, M., Klyuvak Oksana, V., Protsevyat Oksana, S.: Diagnostics of the enterprise export potential. Probl. Econ. (4), 163–170 (2016). Research Centre For Industrial Development Problems of NAS (Kharkiv, Ukraine)

Yalovy, G.K., Bakerenko, N.P.: Conceptual approaches to definition of financial stability of enterprise. Econ. Bull. NTUU "KPI" **8**, 23–28 (2011)

Zachosova, N., Babina, N.: Identification of threats to financial institutions' economic security as an element of the state financial security regulation. Baltic J. Econ. Stud. **4**(3), 80–87 (2018). https://doi.org/10.30525/2256-0742/2018-4-3-80-87

Zakharova, N.Yu.: Methodological approaches to assessing the financial condition of an enterprise. Collection of scientific works of the Tavria State Agrotechnological University (Economic Sciences) **2**(3), 128–133 (2013)

Smart Cities

Open Data and Smart City Initiatives for Digital Transformation in Public Sector in Poland. A Survey

Patrycja Krauze-Maślankowska(✉) (iD)

University of Gdańsk, Gdańsk, Poland
patrycja.krauze-maslankowska@ug.edu.pl

Abstract. The digital transformation influences the dynamics of changes in the public sector. One of the well-known approaches of digital transformation is a smart city concept which become a trend that allows for effective management of urban fabric based on technology. The goal of the article is to present the real readiness of cities in Poland to implement the data-driven approach in the context of the smart city concept. The use of open data in this matter is to implement innovations to improve the quality of life of the inhabitants. To achieve the goal of the article, a survey was conducted in cities in Poland, covering three layers of cities. The survey shows that the data-driven approach increasingly contributes to better city management and improvement of the quality of life of its inhabitants. The article provides statistical data based on the sample of 280 cities in Poland, included in the survey.

Keywords: Smart cities · Big data · Digital transformation · Data-driven approach

1 Introduction

For many years, innovative technological solutions have been used to solve economic, social and environmental problems. The smart city concept is becoming a trend that allows for effective management of urban fabric based on technology. Its main goal is to implement innovations to improve the quality of life of the inhabitants. Cities around the world recognize the value added of implementing the smart city concept into the urban ecosystem. However, making reliable decisions regarding the selection of the best solutions is not possible without access to appropriate information resources. It is because of the analysis of large data sets that cities can minimize the risk of investments, respond to the real needs of residents and implement services tailored to changing conditions. The digital transformation of the public sector is constantly evolving, and in the era of the pandemic, the pace of change has dynamically increased. Many public administrative units are not ready for radical changes in the method of management, information processing and modernization of the technical architecture that will allow for the creation of new services.

© Springer Nature Switzerland AG 2021
S. Wrycza and J. Maślankowski (Eds.): PLAIS EuroSymposium 2021, LNBIP 429, pp. 71–81, 2021.
https://doi.org/10.1007/978-3-030-85893-3_5

The aim of this article is to present the real readiness of cities in Poland to implement the data-driven approach in the context of the smart city concept. With regard to the research carried out to assess the level of implementation of the smart city concept in Polish cities, two research questions were asked. Firstly, do cities in Poland have adequate technological resources to collect, process and analyze data? Secondly, do cities in Poland have human resources capable of drawing conclusions and generating knowledge based on available data in order to increase the effectiveness of decision-making processes? The research method adopted as part of the answers to the questions posed is a comparative city analysis based on a survey conducted in 280 Polish cities. The digital transformation influences the dynamics of changes in the public sector. Data-driven city management and the inclusion of experts and city stakeholders in project teams contribute to building a resilient and sustainable living space.

The first part of this paper focuses on the characteristics of important aspects related to the smart city concept that contribute to the digital transformation of the public sector. This section also covers the value of open data for a suitable transformation. The second and third parts cover the importance of big data technology in managing a smart city, as well as the key competences that the staff responsible for smart city related data analysis should have. The last part covers the results of a survey conducted in Polish cities with regard to their skills and technical tools allowing for the analysis and use of data in the decision-making process.

2   Digital Transformation, Open Data and Smart City Concept

Digital transformation means changes in many aspects of the functioning of both public administrative units and its entire environment. It concerns the change of the management method and the applicable models of information flow. Digital transformation comes from the IT related research, which means its strong technical background (Cruzara et al. 2021). The speed of technology development is very high, which is a challenge for the local municipal administration, which has to look for new ways to respond to emerging expectations. Digital transformation in the area of local government includes such elements as the transformation of organizational structures and the way in which services are provided to residents. In order to improve the effectiveness of the above-mentioned areas, it is necessary to have the ability to analyze the available data, which contribute to the support and improvement of the decision-making process (Battisti 2020).

An important starting point for changes is the awareness of the great value that proper preparation, processing and sharing of data can bring. According to the OECD, the concept of open government data is a philosophy that leads to increased transparency, accountability and the value created by the re-use of government data. Analysis by OECD covering aspects such as:

- disclosure of information relevant to business by the public administration, open access to registers,
- the scale of information related to legal, geographic, meteorological or transport areas,
- data on society.

Based on the research assessing the quality of data shared on national open data platforms, a ranking was created that reflects their openness, usefulness and the possibility of re-use. In the OECD Open, Useful and Re-usable data (OURdata) Index: 2019 ranking, Korea and France were rated with the highest scores. Poland was in 14th place in the comparison with the indication that the quality of the shared data continues to grow (OECD 2019a, b, c, d). The practice followed by Korea boils down to involving stakeholders in the work on selecting the data to be shared, which makes it more useful. The Korean government also offers extensive support for the re-use of data by organizing events promoting the idea of open data and long-term partnerships that affect the high quality of published information (OECD 2019a, b, c, d). France, on the other hand, has introduced an appropriate legal framework that determines the degree of openness of published data. The country is a leader when it comes to data availability. It has feedback mechanisms, which makes the open data platform one of the most advanced. Users can also add their own data sets, which makes the portal even more attractive. As part of the good practices implemented by the French government, it is necessary to indicate the competitions organized in relation to various thematic areas. They are intended to encourage the public to use open data and to combine national data with those provided by other countries, thus contributing to the creation of better analysis and generation of knowledge for a broader understanding of certain phenomena (OECD 2019a, b, c, d). Poland was appreciated in terms of the changes that took place in the developed policies regarding open data. Updating and clarifying the provisions resulted in an improvement in the quality of the data made available, which, however, still does not represent a significant value for society. It was also noted that there was a need to motivate local governments to share information from the collected resources, and it was recommended to take actions to involve stakeholders in the work on defining the scope of useful data (OECD 2019a, b, c, d).

The ability to process, analyse and conclude on the basis of available data is a key element of development in all areas of smart city. Modern technologies, such as the Internet of Things (IoT) or ubiquitous sensors, generate huge amounts of data. Big data mechanisms based on the analysis of structured and unstructured data allow to discover hidden relationships between various phenomena and behaviours. This process, in turn, creates new knowledge, which drives innovation. Each stage leading to the creation of modern solutions that meet social, economic and environmental needs are based on data. However, this process is still underestimated or incomprehensible by local governments in many cities, both in Poland and in the world (Lodato et al. 2021).

Thanks to the growing interest in the implementation of the smart city concept, the willingness to learn about the phenomenon is also growing, as cities manage to effectively respond to changing conditions and the expectations of their residents. The idea of a smart city is mainly about introducing solutions that improve the quality of life of residents. These include, among others, activities such as:

- monitoring installation with the function of image analysis and real-time response to disturbing events, which contributes to the improvement of citizens' safety;
- optimization of road capacity, consisting in directing vehicles to free parking spaces in order to reduce excessive movement around the city in search of a stop. This approach

also influences the choice of an alternative mode of transport when all places are already occupied, which has an additional impact on environmental conditions;

- intelligent street lighting, designed in such a way as to minimize the costs of its modernization, as well as reduce energy consumption;
- changing the way the local public administration functions by tailoring services to the actual needs of residents and by more open and participatory management.

Open government data play a fundamental role in achieving optimal results in relation to the above-mentioned measures. They constitute the basis for ensuring the transparency of public administration, contribute to improving the quality of life, minimizing the risk related to the implementation of investments that will not be useful for residents, provide the opportunity for stakeholders to participate in solving significant problems and encourage the creation of innovation.

However, open data published by public administration units often turn out to be outdated. Additionally, they are placed in a format that prevents their further analysis. This form of things results primarily from the way in which local government units are managed. Information is produced in a manner inconsistent with the applicable standards and published in an incomplete form, for example in the form of a scan of the original document. In this case, it is important that employees have appropriate competences and have a set of tools that allow for easy adjustment of data to a state that will allow them to be freely processed by analytical mechanisms. The next part of the paper is devoted to research on the necessary competences and technologies that local governments should have. It is important from the point of view of effective action in all areas that make up the smart city concept (smart living, smart mobility, smart economy, smart environment, smart people and smart government). Having the right resources also contributes to the ability to collect, process, analyse and generate knowledge from available open government data.

3 Big Data Analysis Technologies as a Factor in the Development of Smart City

Big data sets are now the basis for shaping innovation and improving the quality of life of city dwellers. The very term big data in the context of urban analyses means the links between information from various sources, such as finance, geographic location, social media or the presence of air pollutants. With appropriate tools, methods, good practices and methods of analysis and inference, it is possible to better understand the changing phenomena occurring in cities. Big data technology is of fundamental importance in the process of making rational decisions. This is because it reduces the risk of problems by isolating the key elements of the actions taken that must be ensured for the investment to be successful. In addition, it allows for a comprehensive assessment of residents' expectations and motivates the community to cooperate (Sang et al. 2021).

In smart cities, data from social media or sensors located in the city is generated continuously in real time. However, they are saved in a different frequency and format. Tools such as Apache Spark, Apache Storm and Apache Flink are used to analyse data from various sources and processed in various formats. Apache Spark, compared to

Hadoop, allows faster processing of data streams represented as a Resilient Distributed Datasets (RDD) sequence and an in-memory adaptation function. Apache Storm, on the other hand, is a distributed computing platform that has libraries that enable the use of machine learning. Apache Flink, on the other hand, is an alternative solution to Apache Spark that allows to define features and ensure lower latency. Apache Spark processes batches of data in the form of RDD, while Apache Flink allows you to read successive lines of data in real time (Manjunatha et al. 2020).

Cities striving to implement the smart city concept recognize the importance of data in the development of all areas of operation. The implementation of Internet of Things (IoT) devices, sensors and services for residents in the urban space is associated with the collection of large sets of data. Knowledge of big data analysis methods is a fundamental factor in the development of modern cities (Okwechime et al. 2018). The next part of this paper indicates the key competences that local public administration should pay attention to optimize decision-making processes.

4   Key Competences in Data Management

The demand for personnel with analytical competencies has been high recently. The ability to create new knowledge and forecast possible events affects the effective management and administration of smart city services, including parking systems, intelligent solutions in relation to traffic management or responding to the health and everyday needs of residents. Digital transformation has contributed to new ways of perceiving data. The ability to analyse incoming content in real time leads to the discovery of patterns that allow the creation of innovative data-based solutions. However, mastering the ability to use multiple data sources from urban space is not an easy task even for experienced analysts (Ranjan et al. 2021).

From the point of view of building platforms, products and services to improve the quality of life in the city, an important skill is to understand the mechanisms that characterize city management in an intelligent and sustainable way. People responsible for creating innovative solutions based on data should have not only analytical skills, but also communication with people of different ages and with different expectations, and able to clearly define the goal of the action and the path to achieve it.

In 2019, in Poland, the percentage of enterprises conducting big data analyzes was 8.5%. Large enterprises expressed interest in such services, 28.4% of which conducted analyses of large volumes of data. The largest share of entities using big data was recorded in the information and communication section - 21.4%, and the lowest - in activities related to the real estate market - 5.2% (Information Society 2020). The interestingness in data science was confirmed by the analysis of big data sources collected by CEDEFOP (The European Centre for the Development of Vocational Training). This data source was prepared based on the data posted online in different platforms of job posting, including job advertisements portals, enterprise websites and public employee agencies. In Fig. 1 there is a list of countries in which the skill "accessing and analysing digital data" is the most common.

Figure 1 shows that the highest share of job offers mentioning the skill "accessing and analyzing digital data" is in Luxembourg (58.6%), Ireland (56.7%), Slovak Republic (42.7%), Hungary (40.1%) and Portugal (39.8%).

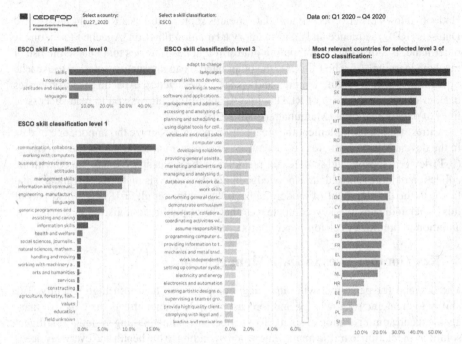

Fig. 1. The skill "accessing and analysing digital data" in online job advertisements by country. Source: CEDEFOP skills OVATE, http://cedefop.europa.eu

5　Characteristics of the Survey and Main Findings

The growing need for public administrative units to have appropriate technical and personal resources allowing for the data collection, processing and analysis in real time, determined the research on the current activities in local governments. The study was aimed at assessing the resources that affect the dynamics of smart city development as well as the prerequisites to digital transformation. The survey was conducted in the territorial self-government units. The survey was conducted from July 1 to August 31, 2020 using the CAWI (Computer-Assisted Web Interview) technique. A randomly selected sample for the study, by the method of simple random sampling, covered 280 cities out of all 940 Polish cities as of June 30, 2020. There were 210 responses, resulting in a return rate of 22% of the total population. In addition, small cities (up to 20,000 inhabitants) account for 70%, medium-sized cities (from 20,000 to 100,000 inhabitants) 21%, and large cities (over 100,000 inhabitants) 9% of all responses, which is proportionally consistent with the real the number of cities in the population for each stratum. The structure of the response against the layers is shown in Table 1.

As can be seen in Table 1, the layer that proportionally contains the largest scale of responses is that corresponding to large cities in Poland. This leads immediately to the conclusion that large cities have more qualified employees who answer all kinds of requests and decide which unit is responsible for providing the necessary information. The conducted research is qualitative in nature, which means that the conclusions are used to assess the state of the surveyed units in relation to selected phenomena.

Table 1. Structure of the study population.

Layers	Sample size	Number of collected questionnaires	Population size	Proportion
Small towns	200	**145**	722	**20%**
Medium-sized cities	60	**47**	179	**26%**
Large cities	20	**18**	39	**46%**
Total	280	**210**	940	**22%**

Source: Own study.

The survey questionnaire consisted of 30 questions. The main goal was to assess the level of implementation of the smart city concept by cities in Poland, which is why the greater part (16 questions) was devoted to this subject. The remaining questions were used to assess the factors important from the point of view of the ability to undertake innovative activities, as well as to indicate barriers and opportunities that cities perceive in relation to development opportunities. In addition, it was asked to indicate whether the unit has an independent position in the organizational structure or a separate unit responsible for the implementation of innovations. The question about the sources of data collected by the organization and the technologies used for their processing and analysis was also taken into account.

The conclusions obtained from the study allow answering the research questions posed. Do cities in Poland have adequate technological resources to collect, process and analyze data?

In this regard, the first step is to define the scale of implementation of various sensors generating data in the urban space. 56% of the surveyed entities (117 cities) declare that they obtain data from sensors generating information on the level of air pollution, noise, the number of vehicles passing through the city or the number of free parking spaces at a given time. Figure 2 shows the percentage of sources from which the data collected by public administration units come.

The summary presented in Fig. 2 shows that most cities in Poland, regardless of their size, collect data from sensors that allow measuring the level of air pollution. The observed dependencies may be explained by the fact that in some cities sensors measuring various types of pollution factors were installed by enterprises as part of free services. However, the data collected in this way is not complete, because the provider of free solutions allows to view the data only in the perspective of a single search. Open data platforms with air pollution resources often fail to analyze the long-term situation. Another frequently occurring area is the monitoring of selective waste collection and water consumption in municipal flats, which is an important aspect from the point of view of the current environmental conditions. Large cities more often than others decide to measure the traffic volume in the city, which may contribute to the improvement of the quality of life of inhabitants and reduction of congestion in the most crowded sections by diverting communication to alternative paths. Additionally, large cities notice the need to monitor connections to the wireless Internet network located in the public life space. This approach is conducive to the development of the creative sector and enables

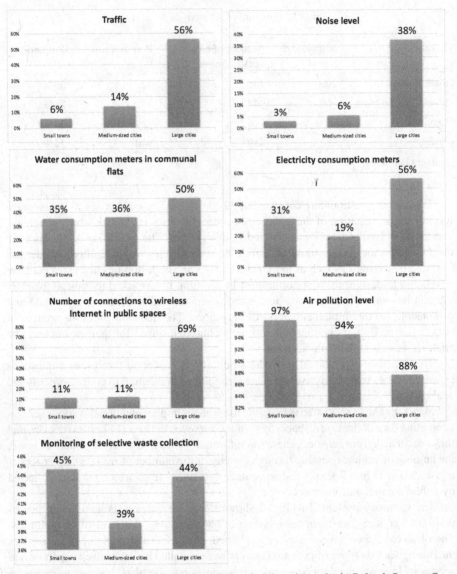

Fig. 2. Types of data sources collected by public administration units in Poland. Source: Own study.

the implementation of innovative urban solutions. In addition to the above-mentioned elements, cities also indicated the use of such types of sensors as monitoring of media consumption in the sports and entertainment hall, monitoring of public transport traffic, monitoring of shared transport vehicle rentals and routes traveled within the city bike system.

In order to effectively use the obtained information, it is necessary to have the tools and knowledge necessary to manage, analyze and interpret city data. Although access to information seems to be getting easier and easier, which should favor the development and implementation of modern technologies, such solutions are still rare in the area of the public sector at the local level. The use of platforms using dynamic and interactive graphical interfaces, 3D models and elements of augmented reality support decision-making processes. At the same time, they contribute to the identification of significant trends. Figure 3 shows the degree of use of modern technological solutions in small, medium and large cities in Poland.

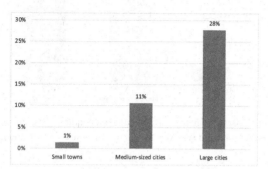

Fig. 3. The use of big data, business intelligence, data warehouses or similar IT systems in data analysis process. Source: Own study.

Out of 210 cities, 117 declared collecting data from sensors. In the total number of cities, only 6% use technology that allows their analysis and interpretation, the remaining part do not use such mechanisms or have no knowledge on this subject. The cost of designing, installing and constantly improving the existing technological solutions that enable the analysis of the collected data, constitute a barrier to entry for smaller cities or those located in less developed regions. An additional limitation may also be the lack of access to good practices in this area, detailed implementation cases and the benefits of adapting integrated technological environments. At the same time, technological instruments developed for a specific project do not allow them to be combined with other solutions, which excludes the possibility of achieving a holistic perspective in relation to the assessment of actions taken (Ruohomaa et al. 2019).

When resolving the question asked at the beginning, most Polish cities have appropriate conditions for the development of innovation. In addition to equipping public space with various types of sensors, local administration also undertakes activities aimed at using the information generated by them.

In addition to the technological resources themselves, which can meet all the assumptions of integrity, adequately qualified and open to new solutions employees are also important. Therefore, the next part presents the results of the study, which are intended to answer the next research question:

Do cities in Poland have human resources competent to draw conclusions and generate knowledge based on available data in order to increase the efficiency of decision-making processes?

Local public administration units in the world more and more often decide to appoint an independent position or a separate organizational unit responsible for the analysis of collected data and the creation of urban innovations. However, this approach is still little practiced in Polish local governments. This trend is presented in Fig. 4.

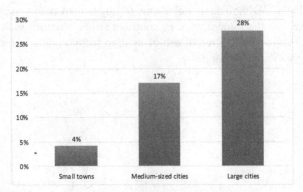

Fig. 4. Units with a separate department or an independent position responsible for the implementation of innovations and services for residents. Source: Own study.

As shown in Fig. 4, only large cities in Poland decide to invest in new jobs that will represent an innovative image of a given area. Collecting appropriate staff is a difficult task for public administration due to the growing competition on the labor market which private enterprises and corporations constitute for municipal offices. Additionally, a big barrier for most individuals is ensuring an adequate level of remuneration that meets the requirements for the expected skills of a given person (David et al. 2019).

The presented results clearly indicate that small and medium-sized cities in Poland do not have adequate human resources that will allow discovering hidden knowledge based on the available data. It also means that they are not able to fully meet the real needs of the inhabitants. In many offices, however, there are people who, apart from their basic scope of duties, engage in work on the creation of innovations. They also often cooperate with the research community and independently develop their skills to transform them into added value for the city.

6 Conclusions

Cities around the world struggle with the growing amount of data stored in their structures. The data-driven approach is a method that increasingly contributes to better city management and improvement of the quality of life of its inhabitants. The dynamically following digital transformation plays an important role in the context of the appropriate adaptation of the technical infrastructure of local governments. It somehow forces the public sector to invest in solutions that generate new knowledge for residents. Creating new knowledge on the basis of available data and transferring it to the management of the office and residents contributes to increasing the transparency of the government (Matheus et al. 2020). In addition, such activities encourage citizens to cooperate with local authorities, which affects mutual trust and understanding. Planning solutions comprehensively tailored to the expectations and social, environmental and economic conditions, however, is related to the possession of appropriate skills by employees employed in local offices. Technical, analytical, communication and cooperation skills are the basic resources that an office should be equipped with in order to make rational decisions based on the collected data. According to the survey, 6% of the cities in Poland which collect data from various types of sensors, do not have an appropriate unit that would adequately manage their data. The growing interest in implementing the smart city concept contributes to the city's achievement of the above-mentioned assumptions. As the presented research results show, cities in Poland have development potential which allows them to adapt to dynamically changing technological conditions. The knowledge and experience of new and long-term employees, often strongly involved in the development of the unit, allows us to meet the expectations of local stakeholders. However, only large cities in Poland are beginning to recognize the growing importance of the data-driven approach and are responding to these changes by investing in human capital with the skills to manage large data sets and the necessary technology.

References

Battisti, D.: The digital transformation of Italy's public sector. JeDEM eJournal eDemocracy Open Gov. **12**(1), 25–39 (2020)

Cruzara, G., Sandri, E.C., Cherobim, A.P., Frega, J.R.: The value at the industry 4.0 and the digital transformation process: evidence from Brazilian small enterprises. Revista Gestão Tecnologia **21**(1), 117–141 (2021)

David, N., McNutt, J.: Building a workforce for smart city governance: challenges and opportunities for the planning and administrative professions. Informatics **6**(4), 47 (2019)

Lodato, T., French, E., Clark, J.: Open government data in the smart city: interoperability, urban knowledge, and linking legacy systems. J. Urban Aff. **43**(4), 586–600 (2021)

Manjunatha, S., Annappa, B.: Real-time big data analytics framework with data blending approach for multiple data sources in smart city applications. Scalable Comput. Pract. Exp. **21**(4), 611–623 (2020)

Matheus, R., Maheshwari, D.: Data science empowering the public: data-driven dashboards for transparent and accountable decision-making in smart cities. Gov. Inf. Q. **37**(3), 101287 (2020)

OECD: OECD OURdata Index: 2019, France (2019a). https://www.oecd.org/gov/digital-govern ment/ourdata-index-france.pdf

OECD: OECD OURdata Index: 2019, Korea (2019b). https://www.oecd.org/gov/digital-govern ment/ourdata-index-korea.pdf

OECD: OECD OURdata Index: 2019, Poland (2019c). https://www.oecd.org/gov/digital-govern ment/ourdata-index-poland.pdf

OECD: Open Government Data (2019d). https://www.oecd.org/gov/digital-government/open-gov ernment-data.htm

Okwechime, E., Duncan, P., Edgar, D.: Big data and smart cities: a public sector organizational learning perspective. IseB **16**(3), 601–625 (2018). https://doi.org/10.1007/s10257-017-0344-0

Ranjan, J., Foropon, C.: Big data analytics in building the competitive intelligence of organizations. Int. J. Inf. Manag. **56**, 102231 (2021)

Ruohomaa, H., Salminen, V., Kunttu, I.: Towards a smart city concept in small cities. Technol. Innov. Manag. Rev. **9**(9) (2019)

Sang, L., Yu, M., Lin, H., Zhang, Z., Jin, R.: Big data, technology capability and construction project quality: a cross-level investigation. Eng. Constr. Architect. Manag. **28**(3) (2021)

CEDEFOP Skills OVATE. http://cedefop.europa.eu. Accessed 16 May 2021

Information society in Poland - report. Statistics Poland, Warsaw (2020)

The Role of Smart Buildings in the Digital Transformation Era: Insight on China

Eleonora Veglianti[1]([✉]), Yaya Li[2], and Marco De Marco[3]

[1] FGES, Université Catholique of Lille, Lille, France
eleonora.veglianti@univ-catholille.fr
[2] School of Finance and Economics, Jiangsu University, Zhenjiang, PR China
[3] Department of Economics, University Uninettuno, Rome, Italy
marco.demarco@uninettunouniversity.net

Abstract. Recently, the Industry 4.0 scenario influences the society toward an important change. The digital transformation pushes the growth of smart cities and of its components such as smart buildings. The latter are playing a crucial role as an innovative concept with creative applications. This paper contributes to the literature gap studying the state of art of smart buildings in a context such as China. Following a qualitative approach, this article uses a new way to better understand smart buildings development strategies to create a smart scenario with building control system evolving toward a higher use of artificial intelligence, neural networks and learning controls.

Keywords: Smart city · Smart building · Urbanization · Societal change · China

1 Introduction

In the current era, the urbanization phenomenon and the growth of the population create new challenges and opportunity to our society. In this context, smart city, smart people, smart buildings, smart mobility are features that reach a predominant position in the academic and professional discussion. Therefore, during the so-called Industry 4.0, our society is going to be characterized by a digital transformation. Specifically, a smart society driven by technology, digital connectivity, new knowledge, and skills as well as innovation impacts the political, social and economic development [1].

Being a smarter society means also to build a new ecosystem with innovative social and technical elements having people interacting with machines [2] with the goal of increasing the quality of life [3]. In line with this, focal points are the energy demand and the waste production [4, 5].The latter features open a vivid debate among scholars as well as practitioners in several countries to better understand its related topics to enhance the quality of life of people as well as to improve the environmental situation.

In this scenario, smart cities in general and smart buildings in particular represent strategic elements to fit new needs and necessities [6, 8]. In this regard, China is facing a crucial revolution since 2009 with a huge increase of smart cities cases due to the urban development [9]. Therefore, new smart cities examples (i.e. Xiong an New Area)

S. Wrycza and J. Maślankowski (Eds.): PLAIS EuroSymposium 2021, LNBIP 429, pp. 82–93, 2021.
https://doi.org/10.1007/978-3-030-85893-3_6

come out reaching a prominent role both in the public and private sphere [10]. Moreover, some areas such as Beijing, Tianjin and Hebei present a radical change toward a smart ecosystem.

In other words, with the continuous development of smart city construction, smart buildings are becoming more and more widespread, representing the future of architecture and of artificial intelligence applications. Thus, for this kind of building, the system equipped with them can realize remote control through external network (such as Internet) which results an important step toward a society that shows its interest in smart technologies.

The relevant literature shows great attention to the concepts related to smart city both at European level as well as globally [7]. However, few scientific contributions focus on smart components such as smart buildings especially investigating on specific contexts outside Europe, specifically in China. Consequently, the research question is the following: Which are the characteristics that mostly influence the development of smart buildings in China?

To answer to this research question, we applied a multi case-based research to carefully study this phenomenon in a narrative approach that occurs with the data collected. The study is on China due to the nationality and the direct experience of the authors in this country. Furthermore, they could read materials in original language, without the need for translation helps to avoid bias and to enhance the tacit knowledge behind the phenomenon under investigation.

This paper fills the literature gap studying the concept of smart building in China to better understand the state-of-art using a case approach. Besides, our aim is to stress the debate on smarter society, in general, and on smart ultimate technologies, in particular, dealing with smart building strategies giving insight on the Chinese context.

The paper is structured as follows: Sect. 2 explores the literature review; Sect. 3 defines the methodology; Sect. 4 analyses the cases under investigations showing the emergent results Sect. 5 presents the discussions on the results and concludes, analysing the limits of the paper and suggesting further works.

2 Literature Review

In the scientific sphere, an umbrella of definitions to analyze the smart city concept emerged [8]. In the last decades, in the relevant scientific literature, several labels appear with some main focuses (Table 1, below).

Recently, the concept of smart city gains great attention also in terms of urban and ecological efficiency. In this context, smart buildings are identified as important components. Therefore, in a society that is going to be smarter and smarter with the increase in size of cities requires the creation of smart buildings [11]. As some scholars suggest these new buildings have to be more efficient, sustainable and long-lasting compared to the past [12]. Others considered smart buildings linked to security and quality of life [13]. Moreover, some authors demonstrated that artificial intelligence impacts the evolution of buildings and the smart transformation of houses allowing the birth of new business models that highlight the intelligence application to architecture [14, 15].

Table 1. Labels in the scientific literature

Labels	Main focus	Literature contribution
Digital city/ Ubiquitous city	Digital, communication, data, networks, broadband	Qi, L., Shaofu, L. (2001). *Research on digital city framework architecture.* IEEE International Conferences on Info-Tech and Info-Net, vol. 1, (pp. 30–36). Proceedings ICII Schuler, D. (2002). *Digital cities and digital citizens.* In: M. Tanabe, P. van den Besselaar, T Anthopoulos, L., Fitsilis, P. (2010, July). *From digital to ubiquitous cities: Defining a common architecture for urban development. In Intelligent Environments* (IE), 2010 Sixth International Conference on (pp. 301–306). IEEE
Intelligent city/ Knowledge city	Information, knowledge creation, human capital	Komninos, N., Pallot, M., Schaffers, H. (2013). *Special issue on smart cities and the future internet in Europe.* Journal of the Knowledge Economy, 4(2), 119–134 Ergazakis, M., Metaxiotis, M., Psarras, J. (2004). *Towards knowledge cities: conceptual analysis and success stories,* Journal of Knowledge Management, Vol. 8 N..5, p.5–15
Green city/ Sustainable city	Green technologies, green energy, environmental issues, sustainability	Gordon, D. (1990). *Green cities: ecologically sound approaches to urban space* (No. 138). Black Rose Books Ltd Batagan, L. (2011). *Smart cities and sustainability models.* Revista de Informatica Economica, 15(3), 80–87
Smart city	Smart people, sustainable growth, technology, quality of life	Giffinger, R., Fertner, C., Kramar, H., Meijers, E. (2007). *City-ranking of European medium-sized cities.* Cent. Reg. Sci. Vienna UT, 1–12 Caragliu, A., Del Bo, C., Nijkamp, P. (2013). *Smart cities in Europe.* Smart cities: governing, modelling and analysing the transition, 173 Dameri, R.P. (2013). *Searching for smart city definition: a comprehensive proposal.* International Journal of Computers & Technology, 11(5), 2544–2551(Council for Innovative Research) Dameri, R.P. (2014). *Comparing Smart and Digital City: Initiatives and Strategies in Amsterdam and Genoa. Are They Digital and/or Smart?* Smart city. How to create Public and Economic Value with High Technology in Urban Space, Springer International

A smart building has to contribute to the sustainability of a smart city. Elements such as micro renewables, as solar and wind, micro energy storage systems and energy consumption controller has to be installed and set up in order to make the building a synergically part of a smart urban environment [16].

In other words, the consumption of energy is one of the main features to take into account to have a smart building with energy consuming devices that are turning on a new light on self-sustainability. In the close future, the role of information and its storage will be more and more strategic presenting a higher number of smart buildings in different cities. This implies that many different information will be stored asking the creation of new artificial intelligence systems. The latter are going to optimize the information storage following low consuming conditions and self-sustainability [17].

At the same time, if smart buildings will be integrated with clouds that use artificial intelligence, new issues related to integrated disaster management will emerge [18]. For example, privacy and cybersecurity represent important aspects in this field [19].

It is necessary to say that every device that uses electricity and, in general, every electrical system can have a crucial role in terms of consumption reduction. Therefore, for instance, the use of lighting technology, information and communication infrastructures, cyber-physical modeling for cooling and ventilation, can bring a fundamental decrease of energy expenditure or an important increase of work efficiency with negligible incremental costs [20].

Several authors discussed the elements which impact self-sustainability of buildings. For example, Moreno [21] suggested a user-centric smart solution to have a lower environmental impact of smart buildings. In addition, Arditi [22] proposed that buildings represent a total energy consumption between 20% to 40%; while, considering only the electricity consumption, buildings are responsible for 70% of total utilization [23]. Another interesting contribution, focused on the Green House Gas produced by buildings all over the world, concludes that houses are responsible for the 40% of the total production [20].

Moreover, other researches tried to define the effect of buildings quality on the life of occupant [24–26]. For example, some provide positive results which allow higher well-being and comfort with a minimum energy expenditure [27]. Other studies focused on the requirements needed to define a smart building. For instance, Minoli [28] identified nine dimensions: server room, office space, HVAC room, cooling system elements, BMS (Building Management System), electrical system, plumbing/water system, common areas and retail areas inside the building. Moreover, the building management system results crucial for the energy management which is composed by sensors, controllers, output devices, communication media and supportive protocols, data analytics.

Additional contributions considered sensing technologies in smart buildings [23, 29, 30]. In this field, an interesting model proposed by Costanzo [31] was composed by three layers: admission control, load balancing, and demand response management. Thus, the interconnection between different devices and dynamics allowed to control the energy consumption and the optimal plan of activities.

Recently, the building control system concept evolved from the application of conventional controllers to controllers based on artificial intelligence, neural networks and learning controls. In this way, there is a better efficiency in terms of energy management [11].

Furthermore, smart building means also smart home. Here, all the devices communicate with each other to guarantee an optimal performance of the daily routine of individuals [32]. Specifically, smart home should be considered as subcategory of smart buildings [33].

In addition, Internet of Things result fundamental for the develop and the application of smart concepts in buildings and houses. The challenges that has been identified for the design of smart environment through Internet of Things could be summarized in availability, reliability, interoperability, scalability, performance management, security and privacy, big data analytics, cloud services, and smart devices design [33]. For instance, cloud computing is one of the instruments through which this kind of challenges are faced [34]. Thus, as the relevant literature suggests, Internet of Things is critical in the development and application of smart concepts in buildings with cloud computing playing a crucial role [34].

To conclude, given the growing interest on energy management in smart cities as confirmed by the literature (i.e. [16]), smart buildings can provide different interrelated services allowing smart cities to meet the increasing consumer demand for a more efficient situation in an urban agglomeration context.

3 Research Methodology

This article presents a case study approach to allow the exploration and understanding of a complex issue such as smart building in China. This method is useful to have a deep analysis of the social issues relevant in the field in comment.

The aim is to deep study the data within a specific context, as Yin [35] said "the case study research method is an empirical inquiry that investigates a contemporary phenomenon within its real-life context; when the boundaries between phenomenon and context are not clearly evident; and in which multiple sources of evidence are used." This method allows the analysis of the data within the context and within the situation in which the activity takes place [35].

The presents work uses a multiple-case design [36, 37]. Moreover, considering the relevant literature in this field and the purpose of this contribution, we adopted a descriptive approach that allow a better understanding of the phenomenon under investigation in a narrative way that occur within the data under examination [38]. Following this relevant literature, the present work defines a multi-case method to describe the *state-of-art* of smart building in the Chinese context, as described in the next paragraphs.

China is the context of the present paper as it is a pioneer in several technological applications. Moreover, China plays an important role in this environment battle. This country has invested a lot in smart cities since 2009 to face the urban development [9]. It is increasingly focused on a sustainable urbanization [39]. Numbers immediately show this trend with around 300 smart cities pilot projects [40–42].

Therefore, the interest in the development of smart cities in general and of its components as smart buildings in particular is clear. For example, according to the announcement of national standards of the People's Republic of China (no. 13 of 2019), there is an important approval for the "general technical requirements for comprehensive service platform of smart city construction and residential area (standard number GB/T 38237-2019) [43]. The standard has been in force since May 1, 2020. Moreover, the comprehensive service platform of building and residential area is the core of smart society management, service and operation.

The formulation of this standard is conducive to the formation of additional connectivity between different smart society terminal systems, service business and management, and constitutes a flexible and efficient technical framework system of smart community. Therefore, it represents an important guide to promote the future construction and operation of a smart society in China and to better support the development of smart city. Here, focusing on different elements such as smart buildings and smart homes, the national smart standards commission confirms an increasing interest on these themes (Connectivity partners conference, Shenzhen, 2019).

As a consequence, China is a breeding ground in this field given the interest in the construction of a smart ecosphere following peculiar standards [44].

Given this prosperous context, we identified through a deep research the cases that are significant for the Chinese community. The choice was driven mainly by three reasons. Firstly, the availability of information which were gathered from different websites (i.e. China Association of Urban Science, Smart City Innovation Solutions Research Center) as well as blogs, news that the authors were able to reach and read directly in Chinese to avoid the translation bias. Secondly, the tacit knowledge and the personal experiences in China of the authors help the research to find additional insights in the selection of the cases as pioneer examples of energy efficiency in China. Finally, the access to recent scientific contributions in Chinese on smart building drives the choice to better understand the *state-of-art* of smart building in China.

4 Emerging Results

4.1 The Case of Tencent Binhai Building

Tencent Binhai Building is located in the science and Technology Park of Nanshan District, Shenzhen with a total investment of 1.8 billion yuan (Source: Sohu). Each floor is designed in loft style and connected by three "belts", covering an area of 18.650 square meters and a building area of about 350.000 square meters.

This building represents one of the main pioneer examples of smart building in China, especially in terms of Internet of Things and artificial intelligence promoted by a company. The development of such types of building is capturing increasing attention in China (i.e. [45, 46]) In fact, this case uses Internet of Things and artificial intelligence technology, thus it integrates digitization and intelligence.

This case results innovative for different reasons. Firstly, it is a building created to improve the environment protection; in fact, the concept of green environmental protection runs through the whole process of its construction. Secondly, it is a Tencent building thus created by a giant Chinese company to increase the quality of the employees. In line with this, for example, the employees do not need to swipe their cards to enter the office area, but directly scan their faces. The elevator system is integrated into the building Internet of things; therefore, people can book elevators and reach the floors using Wechat. The latter is used also to have the lighting control in the office area.

Then, it is possible to achieve a precise indoor positioning of one meter. In addition, QQ account supports intelligent car searching navigation system. In terms of architecture and lighting efficiency, the wall is designed according to the local sunshine trend of Shenzhen; therefore, the full floor glass window has its own sunshade system. Another interesting element to be efficient is that its swimming pool can recover the heat from the room to maintain a constant temperature.

From our analysis, Tencent Binhai building represents a vertical ecological community that improves the quality of life of the employees in line with the goal of improving the efficiency in an ecological lens.

4.2   The Case of "0+ Cabin" Building

In the Zhong Xin Tianjin city, smart building was completed in Huifengxi Smart Energy Town. The latter represents one of the main pioneer examples of smart building in China, as it represents a zero-energy building. This case is also interesting as it is an important part of the strategic cooperation agreement between the State Grid Corporation of China and the city of Tianjin. Specifically, the building, named "0+ Cabin", achieves full energy self-sufficiency. Moreover, the 135 square meter "0+ hut" makes the most of available space such as roof and pavement, laying 60 photovoltaic panels.

Specifically, under good lighting conditions, it can generate 60 degrees of electricity every day, fully achieving self-sufficiency in building energy, and realizing a surplus electricity, becoming a regional distributed power supply. Hence, it is equipped with a zero-energy building operation management system and a 40-kilowatt-hour energy storage facility, the photovoltaic energy can be stored and discharged to the building's electrical equipment at night and in the case of insufficient light. In addition, for instance, in the case of surplus electricity, the water heater and washing machine can be started and adjusted at the right time, so as to effectively allocate the electricity and automatically complete the scheduled housework.

On the management platform of zero energy building operation system, information such as photovoltaic power generation, energy storage and charging amount, load electricity consumption, as well as environmental conditions such as temperature and humidity are clearly displayed. Therefore, through application modes settings such as intelligent, comfortable, energy saving, personality and night, the energy demand of different users can be met.

In other words, it results an innovative example because with the use of machine learning, big data analysis and other technologies, the system can automatically adjust the energy consumption equipment in different areas of the building. For instance, it adopts advanced micro grid to reduce loss and achieve energy conversion efficiency of more than 95%.

In conclusion, our results suggest that the "0+ Cabin" is an important case in the Chinese scenario created by the cooperation between the central government and the local municipality in Tianjin having as main goal the energy optimization with a real-time energy and electricity analysis that provides users a crucial service about their energy consumption.

4.3 The Case of Nanjing Jiangbei New District Civic Center

The case of Nanjing Jiangbei New District Civic Center is relevant as it represents the China's first 3D printed assembly type smart building. The project adopts the intelligent construction technology of prefabricated 3D printing exterior wall plus prefabricated fair-face concrete interior decoration and virtual and reality coupling project management system. This is a very innovative building also in terms of lighting system; it is the first case of 360° projection imaging for an integral circular building in China. In particular, the traditional lamps are abandoned, and customized lamps composed of multiple independent units are adopted to achieve the concealment, anti-glare, unity and beauty of the lamps. As night falls, elements such as plum, bamboo, mountain, water and city are displayed on the big screen of Jiangbei Civic Center.

In addition, it presents the highest unsupported escalator in China. Specifically, these are the highest public heavy-duty escalators installed without intermediate support in China, with a lifting height of 19.02 m and a horizontal span of 41.879 m of trusses (Source: Sohu). Finally, the project adopts the largest external wall shading system in China. The above and lower round shuttle-shaped louver system constitutes a music box form, which perfectly integrates the architectural facade effect and the shading system.

Thus, our results highlight that the Nanjing Jiangbei New District Civic Center results an important case in the Chinese scenario as it brings a mix of architectural harmony setting and energy optimization efficiency.

5 Discussion and Conclusions

Currently, China is an important player in the sustainable urban planning given the huge population and the urbanization issues. In this scenario, smart cities, in general, and smart buildings, in particular, represent key elements for the management of a sustainable urbanization plan. From the analysis, it is clear that the establishment and the development of smart buildings in China is confirmed by a number of project initiatives that are wide spreading in all the country.

In order to answer to our RQ, we found that actions and inputs to promote the development of smart buildings in China are implemented with common features as well as with some specific elements. Regarding the common elements, the cases under investigation highlight that these buildings have to be more efficient, sustainable and long-lasting which is in line with the literature of reference (i.e. [12, 16, 20]). This focus on the efficiency and sustainability is at the base of smart building projects in China. Another important features that emerged from our findings is the attention given to the quality of life. As also some scholars argued (i.e. [13]), citizens as well as employees that "live" the smart building should have an improvement in terms of life quality standards. Therefore, a smart building should have as one of its goal the enhancement of people life standard quality. Moreover, much more smart buildings are efficient and, at the same time, are integrated with architectural effects to be more harmonious with the external environment.

In addition, the specific features that come out from our analysis is that some initiatives are managed by large technological companies such as the Tencent Binhai Building case demonstrates. While others are driven by the central government in cooperation with local entities such as the city of Tianjin with the 0+ Cabin example.

This analysis contributes to shed a light on the common patterns as well as on specificities of different smart building projects in China to define a *fil rouge* of the features that influence and drive their development in such a unique context. In other words, these cases focused on innovative technical and technological standards that want to improve the environmental situation with also the aim at creating a space that enhance the quality of life of the people. Therefore, even if each example has its own specificities it show that a clear planning and construction based on an advanced intelligent layout to respond to the current needs of reaching a better quality for people's life as well as for the overall environment.

In conclusion, the three cases under investigation confirmed the relevant literature suggesting that the building control system concept evolved toward a higher use of artificial intelligence and Internet of Things. In this way, there is both a higher efficiency in terms of energy management as well as higher standards for the human beings. Thus, the digital transformation and the evolution of the technological tools impact the society towards smarter building and, as a consequence, toward a smarter society.

In China, as mentioned above, smart buildings are expected to provide and drive the required change toward a smarter society. Artificial intelligence, integrated systems, efficiency settings, innovative services and 3D printings are increasingly adopted.

These findings have significant implications for further scholarly research alike and for practitioners. Additional studies could present an in-depth investigation of smart building comparing different countries and different economies. Despite the importance of this theme in the current society, the article purpose is not to reach a theoretical framework of the phenomenon or to provide implications for the political sphere. However, this paper opens new debates about the *state of art* of smart buildings in China that need further investigations and discussions.

Acknowledgments. This study was supported by National Nature Science Foundation of China Project (No. 71704069).

References

1. Manda, M.I., Backhouse, J.: Towards a "smart society" through a connected and smart citizenry in South Africa: a review of the national broadband strategy and policy. In: Scholl, H.J., et al. (eds.) EGOVIS 2016. LNCS, vol. 9820, pp. 228–240. Springer, Cham (2016). https://doi.org/10.1007/978-3-319-44421-5_18

2. Scekic, O., Nastic, S., Dustdar, S.: Blockchain-supported smart city platform for social value co-creation and exchange. IEEE Internet Comput. **23**(1), 19–28 (2019). https://doi.org/10.1109/MIC.2018.2881518

3. Levy, C., Wong, D.: Toward a Smart Society. Big Innovation Centre, London (2014)

4. Cohen, B.: Urbanization, city growth, and the new United Nations development agenda. Cornerstone Off. J. World Coal Ind. **3**(2), 4–7 (2015)
5. Nam, T., Pardo, T.A.: Smart city as urban innovation: focusing on management, policy, and context. In: Proceedings of the 5th International Conference on Theory and Practice of Electronic Governance, pp. 185–194, ACM (2011)
6. Shapiro, J.M.: Smart cities: quality of life, productivity, and the growth effects of human capital. Rev. Econ. Stat. **88**(2), 324–335 (2006)
7. Caragliu, A., Del Bo, C., Nijkamp, P.: 10 Smart cities in Europe. In: Smart Cities: Governing, Modelling and Analysing the Transition, pp. 173 (2013)
8. Dameri. R.P., Cocchia, A.: La valutazione socio-economica della città digitale [Transl. The socio-economic evaluation of the digital city]. In: Liguria, C.T.I. (eds), La città digitale. Sistema nervoso della smart city [Transl. The Digital City. Nervous System of the Smart City], Franco Angeli, Milano (2014)
9. Li, S., Xu, L.D., Zhao, S.: The Internet of Things: a survey. Inf. Syst. Front. **17**(2), 243–259 (2014)
10. Veglianti, E., Magnaghi, E., De Marco, M., Li, Y.: Smart city in China: the state of art of Xiong an new area. In: Magnaghi, E., Flambard, V., Mancini, D., Jacques, J., Gouvy, N. (eds.) Organizing Smart Buildings and Cities. LNISO, vol. 36, pp. 81–97. Springer, Cham (2021). https://doi.org/10.1007/978-3-030-60607-7_6
11. Shaikh, P., Bin Mohd Nor, N., Nallagownden, P., Elamvazuthi, I., Ibrahim, T.: A review on optimized control systems for building energy and comfort management of smart sustainable buildings. Renew. Sustain. Energy Rev. **34**, 409–429 (2014)
12. Tanda, A., De Marco, A.: How do smart building projects define and deliver value? A classification of business modelling characteristics to support design and development. In: International Conferences ICT, Society, and Human Beings. pp. 267–277. Politecnico di Torino, Torino (2019)
13. Gurgen, L., Gunalp, O., Benazzouz, Y., Gallissot, M.: Self-aware cyber-physical systems and applications in smart buildings and cities. In: Design, Automation & Test in Europe Conference & Exhibition (2013)
14. Mahizhnan, A.: Smart cities: the Singapore case. Cities **16**(1), 13–18 (1999)
15. Xu, Y., Ahokangas, P., Turunen, M., Mäntymäki, M., Heikkilä, J.: Platform-based business models: insights from an emerging AI-enabled smart building ecosystem. Electronics **8**(1150), 1–19 (2019)
16. Morvaj, B., Lugaric, L., Krajcar, S.: Demonstrating smart buildings and smart grid features in a smart energy city. In: Proceedings of 3rd International Youth Conference on Energetics, pp. 1–8. IYCE (2011)
17. Kumar, N., Vasilakos, A., Rodrigues, J.: A multi-tenant cloud-based DC nano grid for self-sustained smart buildings in smart cities. IEEE Commun. Mag. **55**(3), 14–21 (2017)
18. Asimakopoulou, E., Bessis, N.: Buildings and crowds: forming smart cities for more effective disaster management. In: Fifth International Conference on Innovative Mobile and Internet Services in Ubiquitous Computing (2011)
19. Khatoun, R., Zeadally, S.: Cybersecurity and privacy solutions in smart cities. IEEE Commun. Mag. **55**(3), 51–59 (2017)
20. Kleissl, J., Agarwal, Y.: Cyber-physical energy systems: focus on smart buildings. In: Design Automation Conference (DAC), Anaheim, CA, USA (2010)
21. Moreno, M., Zamora, M., Skarmeta, A.: User-centric smart buildings for energy sustainable smart cities. Trans. Emerg. Telecommun. Technol. **25**(1), 41–55 (2013)
22. Arditi, D.: Assessing the smartness of buildings. Facilities **33**(9), 553–572 (2015)
23. Weng, T., Agarwal, Y.: From buildings to smart buildings – sensing and actuation to improve energy efficiency. IEEE Des. Test Comput. **29**(4), 36–44 (2012)

24. Wilner, D.: The Housing Environment and Family Life: A Longitudinal Study of the Effects of Housing on Morbidity and Mental Health. Johns Hopkins Press, Baltimore (1962)
25. Elton, P., Parker, J.: A prospective randomized trial of the value of rehousing on the grounds of mental health. J. Chronic Dis. **39**, 221–227 (1986)
26. Holopainen, R.: Comfort assessment in the context of sustainable buildings: comparison of simplified and detailed human thermal sensation methods. Build. Environ. **71**(1), 60–70 (2014)
27. Wang, Z., Wang, L., Dounis, A., Yang, R.: Multi-agent control system with information fusion-based comfort modelfor smart buildings. Appl. Energy **99**, 247–254 (2012)
28. Minoli, D., Sohraby, K., Occhiogrosso, B.: IoT considerations, requirements, and architectures for smart buildings—energy optimization and next-generation building management systems. IEEE Internet Things J. **4**(1), 269–283 (2017)
29. Froehlich, J., Larson, E., Gupta, S., Cohn, G., Reynolds, M., Patel, S.: Disaggregated end-use energy sensing for the smart grid. IEEE Pervasive Comput. **10**, 28–39 (2011)
30. Kumar, A., Singh, A., Kumar, A., Kumar Singh, M., Mahanta, P., Mukhopadhyay, S.: Sensing technologies for monitoring intelligent buildings: a review. IEEE Sens. J. **18**(12), 4847–4860 (2018)
31. Costanzo, G., Zhu, G., Anjos, M., Savard, G.: A system architecture for autonomous demand side load management in smart buildings. IEEE Trans. Smart Grid **3**(4), 2157–2165 (2012)
32. Ghayvat, H., Mukhopadhyay, S., Gui, X., Suryadevara, N.: WSN- and IOT-based smart homes and their extension to smart buildings. Sensors **15**, 10350–10379 (2015)
33. Khajenasiri, I., Estebsari, A., Verhelst, M., Gielen, G.: A review on Internet of Things solutions for intelligent energy control in buildings for smart city applications. Energy Procedia **111**, 770–779 (2017)
34. Plageras, A., Psannins, K., Stergiou, C., Wang, H., Gupta, B.: Efficient IoT-based sensor BIG data collection – processing and analysis in smart buildings. Futur. Gener. Comput. Syst. **82**, 349–357 (2018)
35. Yin, R.K.: Case Study Research: Design and Methods. Sage Publications, Beverly Hills (1984)
36. Yin, R.: Case Study Research: Design and Methods, 2nd edn. Sage Publishing, Beverly Hills (1994)
37. Campbell, D.: Degrees of freedom and the case study. Comp. Pol. Stud. **8**, 178–185 (1975)
38. McDonough, J., McDonough, S.: Research Methods for English Language Teachers. Arnold, London (1997)
39. Tan, Y., Xu, H., Zhang, X.: Sustainable urbanization in China: a comprehensive literature review. Cities **55**, 82–93 (2016)
40. EU Parliament: Mapping Smart Cities in the EU (2014)
41. EU Commission: The EU Explained: Digital Agenda for Europe. Publications Office of the European Union, Luxembourg (2014)
42. EU-China Smart and Green City Cooperation: Comparative study of smart cities in Europe and China. In: White Paper (Draft), EU-China Policy Dialogues Support Facilities II, March 2014
43. General Technical Requirements: National Standards officially released. Intel. Building Smart City 2019(12), 12 (2019) [4]李想.《智慧城市建筑及居住区综合服务平台通用技术要求》国家标准正式发布[J].智能建筑与智慧城市,2019(12):12.
44. Li, X.: Smart city architecture and residential comprehensive service platform. General technical requirements> national standards officially released. Intel. Building Smart City, 2019(12), 12 (2019) 李想.《智慧城市建筑及居住区综合服务平台通用技术要求》国家标准正式发布[J].智能建筑与智慧城市,2019(12), 12 (2019)

45. Yang, S., Yang, D.: Application of artificial intelligence in intelligent building. Intel. Building Smart City, 2(03), 30–33 (2020)杨诗冬,杨邓文萍.人工智能在智慧建筑中的应用[J].智能建筑与智慧城市, 2(03), 30–33 (2020)
46. Wang, H., Han, C., Li, D., Li, H.: The latest development and application of AIoT technology in building automation system of green intelligent buiding. J. Central China Normal Univ. 55(01), 52–60 (2021) 王宏,韩晨,李丹丹,李红涛.AIoT技术在绿色智能建筑楼宇自控系统中的最新发展和应用探究[J].华中师范大学学报(自然科学版) 55(01), 52–60 (2021)

Digital Education

Digital Education and Information Security in Obstetric Students in COVID-19 Pandemic Times in Peru

Augusto Felix Olaza-Maguiña(✉) and Yuliana Mercedes De La Cruz-Ramirez

Universidad Nacional Santiago Antúnez de Mayolo, Centenario 200, Huaraz 02002, Peru
{aolazam,ydelacruzr}@unasam.edu.pe

Abstract. The objective was to determine the relationship between the digital education topics addressed during the learning of obstetric students and the perception of the security of their information during the virtual education developed because of COVID-19 at the Santiago Antúnez of Mayolo National University (UNASAM) (Huaraz-Peru). It was a cross-sectional research carried out with 179 students, using a previously validated online questionnaire. The SPSS program and the Chi square test ($p < 0.05$) were used. The majority of obstetric students (64.8%) denied having developed the digital education topics consulted during the virtual learning sessions, mainly in the establishment of secure keys (83.8%) and the protection of confidentiality (82.7%); also evidencing a majority perception of insecurity regarding the safeguarding of their academic information on the Internet. It was concluded that there is a statistically significant relationship between the insufficient explanation of digital education topics during the learning of obstetric students and the high perception of insecurity of their information during virtual education developed because of COVID-19.

Keywords: Digital education · Information security · Obstetrics · COVID-19

1 Introduction

Digital education has become in recent years a very important aspect [1, 2], product of the increasingly frequent access to the Internet not only for entertainment purposes, but also for the development of various activities in the productive sector and academic, with the consequent risk of insecurity of data shared through various applications [3, 4].

In this sense, information security and people's digital education to guarantee said protection has been applied with greater emphasis on the training of professionals in areas related to computer science [1, 5], leaving practically aside to university students from other professional careers, who, due to the characteristics of their own academic activities, also maintain contact with other users through virtual platforms and applications [6–8].

In correspondence to what was mentioned in the preceding paragraphs, most of the studies have been oriented to the determination of computer solutions through the use of software that ensures the security of information in university students [1, 9], finding

S. Wrycza and J. Maślankowski (Eds.): PLAIS EuroSymposium 2021, LNBIP 429, pp. 97–107, 2021.
https://doi.org/10.1007/978-3-030-85893-3_7

very few publications with students from the medical area [6, 10, 11], who, in addition to sharing the same risks of insecurity of their data, have a special need to receive a digital education that allows them to protect cybersecurity [5, 7, 10, 12] and confidentiality of his patients [6, 13].

On the other hand, the COVID-19 pandemic has revealed profound deficiencies with respect to the development of digital education in Peru at all educational levels [14], including university students from health sciences careers such as obstetrics, who for many years have not had access to virtual educational sessions, having had to adapt to this learning process without proper preparation and in conditions of high digital divide [15, 16].

The situation described above had not been previously addressed in research related to digital education and information security in obstetric students, especially in vulnerable circumstances such as those evidenced at the Santiago Antúnez de Mayolo National University (UNASAM), not only because of its remote geographical location (3,052 m above sea level), but also because of the type of state management that limits the comprehensive training of students due to lack of resources.

As a result of the analysis of the reality described above, the following research questions were posed:

What are the digital education topics addressed during obstetric students' learning in COVID-19 pandemic times?

What is the perception of the security of their information that obstetric students have during the virtual education developed because of COVID-19?

What relationship exists between the digital education topics addressed during the learning of obstetric students and the perception of the security of their information at UNASAM?

The hypotheses of the study aimed at anticipating limitations in the digital education of obstetric students together with the perception of problems regarding the security of their information, allowed to establish as the main objective of the research to determine the relationship between digital education topics addressed during the learning of obstetric students and the perception of the security of their information during the virtual education developed because of COVID-19 at UNASAM.

2 Research Methodology

2.1 Research Design and Population Under Study

The research design was cross-sectional correlational and the population under study was made up of all obstetric students from UNASAM (204), who met the inclusion requirements of being enrolled in the 2020-II semester, being of legal age (≥ 18 years) and have permanent attendance at virtual classes, considering as an exclusion criterion non-attendance greater than 30%.

Of the total population under study, 97.5% of students participated (199), of which a part of them participated only in the pilot test to assess the reliability of the applied questionnaire (20), while the other students (179) participated in the final data collection.

2.2 Variables

Demographic Characteristics. The demographic variables evaluated were age, sex and place of residence (see the Appendix section).

Topics of Information Security Education. As a result of the review of previous research [2, 3, 6], it was determined the inclusion of dichotomous questions related to the teaching of the basic topics of digital education, which have been detailed in the Table section, having preferred the application of this type of questions to ensure definitive and firm answers from the students for the complete filling of the questionnaire, because the online application of the questionnaire did not allow us to clarify and/or explain the content of the questions.

Perception of Information Security. To evaluate this perception, the general experience of the students during the 2020-II semester in the middle of the COVID-19 pandemic was considered (Appendix section).

2.3 Data Collection Procedure

An online questionnaire was applied between March and April 2021, for which a virtual form was used, the link of which was sent to the students' email. It should be clarified that the applied questionnaire corresponded to the final corrected version of said data collection instrument, which was evaluated for its validity through expert judgment, whose application of Kendall's concordance test demonstrated its validity ($p < 0.001$).

As previously mentioned, the reliability of the questionnaire was also evaluated through a pilot test, obtaining as result a value of 0.862 with the Cronbach's alpha index.

2.4 Statistical Analysis

The statistical program SPSS version 23.0 and the descriptive analysis of the results were applied. The Chi square test ($p < 0.05$) was used to determine the relationship between the digital education topics addressed during the learning of obstetric students and the perception of the security of their information during the virtual education developed because of COVID-19.

2.5 Ethical Considerations

Prior to the application of the questionnaire, the virtual filling of an informed consent form was requested, in which it was clarified that participation was totally voluntary, obtaining the respective authorization for the execution of the study by the Ethics Committee of UNASAM.

Table 1. Demographic characteristics.

Variable	n	%
Age:		
- 18–29 years	168	93.9
- ≥ 30 years	11	6.1
Sex:		
- Woman	155	86.6
- Man	24	13.4
Place of residence:		
- Rural zone	95	53.1
- Urban zone	84	46.9

3 Results

Table 1 shows the results of the demographic characteristics of the 179 obstetric students who participated in the final data collection, being the majority female students (86.6%) residents of the rural zone (53.1%).

The highest percentage of obstetric students considered the safeguarding of their academic information as unsafe (64.8%), as evidenced in Table 2, highlighting among them the non-teaching of the following topics on the basic principles of digital education: Basic definitions (60.9%), relevance of the protection of confidentiality (59.2%) and methods of safeguarding confidentiality (61.4%). Likewise, a statistically significant relationship was found between not teaching the aforementioned topics and students' perception of the security of their information ($p < 0.05$).

Regarding the application of secure keys, the largest proportion of obstetric students who considered the safeguarding of their information insecure (Table 3), denied having developed the following aspects during the virtual learning sessions at UNASAM: Methods for establishing secure keys (60.3%) and use of secure keys (59.8%); results that showed that the perception of information security has a statistically significant relationship with the non-teaching of aspects related to secure keys ($p < 0.05$).

Table 4 shows that in the majority of cases of perception of insecurity regarding the protection of information, students were not explained about the safe use of social networks (57.0%) and computer applications (57.5%) during the development of their academic activities, aspects that in turn showed a statistically significant relationship with the high perception of insecurity ($p < 0.05$).

4 Discussion

The high percentage found in the present research regarding the non-teaching of digital education subjects to obstetric students, has similarities with the deficiencies reported in previous studies [3, 7]. This reality is even more evident in those students whose

Table 2. Basic principles of digital education according to the perception of information security.

Principles	Perception of information security						Total		p-value*
	High security		Medium security		Low security				
	n	%	n	%	n	%	n	%	
Basic definitions:									
- Yes	16	9.0	11	6.1	7	3.9	34	19.0	< 0.001
- No	2	1.1	34	19.0	109	60.9	145	81.0	
Relevance of the protection of confidentiality:									
- Yes	15	8.4	12	6.7	10	5.6	37	20.7	< 0.001
- No	3	1.7	33	18.4	106	59.2	142	79.3	
Methods of safeguarding confidentiality:									
- Yes	17	9.5	8	4.5	6	3.4	31	17.3	< 0.001
- No	1	0.6	37	20.6	110	61.4	148	82.7	
Total	18	10.1	45	25.1	116	64.8	179	100	

* Chi square test.

Table 3. Secure keys according to the perception of information security.

Secure keys	Perception of information security						Total		p-value*
	High security		Medium security		Low security				
	n	%	n	%	n	%	n	%	
Methods for establishing secure keys:									
- Yes	14	7.8	7	3.9	8	4.5	29	16.2	< 0.001
- No	4	2.3	38	21.2	108	60.3	150	83.8	
Use of secure keys:									
- Yes	13	7.3	10	5.6	9	5.0	32	17.9	< 0.001
- No	5	2.8	35	19.5	107	59.8	147	82.1	
Total	18	10.1	45	25.1	116	64.8	179	100	

* Chi square test.

study plan prioritizes professional training, without considering the development of other competencies that, as a result of technological advances, will also have profound importance for the good performance of their work functions, such as example the filling of digital medical records in medical careers [6, 17].

The COVID-19 pandemic has made more noticeable the consequences of the limited digital education that so far has been developed in university students in Peru and other

Table 4. Safe use of social networks and computer applications according to the perception of information security.

Safe use	Perception of information security						Total		p-value*
	High security		Medium security		Low security				
	n	%	n	%	n	%	n	%	
Precautions in the use of social networks:									
- Yes	17	9.5	9	5.0	14	7.8	40	22.3	< 0.001
- No	1	0.6	36	20.1	102	57.0	139	77.7	
Precautions in the use of computer applications:									
- Yes	12	6.7	11	6.1	13	7.3	36	20.1	< 0.001
- No	6	3.4	34	19.0	103	57.5	143	79.9	
Total	18	10.1	45	25.1	116	64.8	179	100	

* Chi square test.

countries with similar characteristics [17–19]. In the present study, the explanation of the results evidenced in the obstetrics students of UNASAM, would be related to multiple factors such as the non-consideration of digital education in the virtual curriculum, as well as the scarce training of teachers in this subject, despite having been performing virtual teaching tasks for more than 1 year.

In correspondence with the aforementioned, a scant explanation of the basic principles of digital education was determined, such as the basic definitions, relevance and methods of protecting confidentiality. These results are due to the fact that all the subjects are in charge of obstetric professionals, and no specialist in computer science has been considered, which, as highlighted in other research, does not allow the promotion of a culture of care for privacy and the protection of personal and professional data [1, 6], culture that other countries are beginning to forge from the most elementary educational levels [3, 20].

On the other hand, various studies have also found problems regarding the training of the general population on the establishment and use of secure keys [3, 21, 22]. This problem would not only have increased the theft of sensitive personal and financial information for users, but also the loss of academic information among students, including confidential data from clinical cases shared for strictly educational purposes [6, 13, 23]. It should be noted that this issue of digital education on secure keys should be approached in a transversal way in all aspects of people's daily lives, even in the use of apparently non-risky mobile applications, which have increased their use during the COVID-19 pandemic.

Another aspect related to the aforementioned findings is the safe use of social networks and computer applications, whose teaching was also deficient among obstetric students despite its importance, even more if it is taken into account that due to the lack of implementation at UNASAM with all the resources and digital tools, many teachers have had to make use of these social networks for the development of their virtual learning sessions. It is therefore necessary to teach this subject of digital education, which, as concluded in other investigations [1, 3, 6, 10] represents the application of easily accessible techniques [7, 24], but if the respective preventive measures are not taken, it can result in extremely dangerous tools for the safety of the students.

It is important to mention that this research provides interesting results regarding the perception of insecurity that students have regarding the protection of their information on the Internet, especially during the fulfillment of their academic activities, a perception that according to the results presented is related to insufficient development of digital education topics at UNASAM. In this regard, there is coincidence with other studies [7, 13], in which mention is also made of the impact that the COVID-19 pandemic has had on the increase in the feeling of insecurity on the Internet, with very particular consequences among students of health sciences, whose majority ignorance of digital education is very worrying in Peru, a country where the effects of the pandemic are still being suffered, among other things due to the slow vaccination carried out by the Peruvian authorities.

5 Limitations

Like all research, there were some limitations that it is important to state and recognize with the intention of allowing an objective analysis of the results found. In this sense, one of the most important limitations corresponded to the restricted scope of the study field, which only included obstetrics students from UNASAM, with which there could be differences with students from other geographical regions of Peru and the world. This limitation is explained by the fact that this study arose as a research line initiative in Peru, specifically in health sciences careers, where to date no research had been done in this regard, so given the lack of antecedents, the findings disclosed can be taken into account to expand new areas of study in this regard.

Other limitations were referred to the non-inclusion of additional aspects on digital education and information security in the questionnaire used, which was due, on the one hand, to the fact that, as stated above, it was a first research initiative in the obstetric career, as well as that there were many inconveniences for a more detailed and long-term study of the variables, as a result of the declaration of the state of health emergency throughout Peru due to the pandemic.

6 Conclusions and Future Steps

6.1 Conclusions

It is concluded that there is a relationship between the insufficient explanation of digital education topics during the learning of obstetric students and the high perception of insecurity of their information during virtual education developed because of COVID-19 at UNASAM. In this sense, university authorities should consider this situation for the implementation of plans to improve the curriculum of this and other professional careers at UNASAM, with the consequent evaluation of the situation in which students from other educational institutions find themselves.

6.2 Future Steps

It is important to carry out new research aimed at the analysis of the variables addressed in this study, but not only in students of other professional careers, but also from the point of view of other educational actors such as teachers and management personnel, which would provide greater scope regarding the importance and content that should be developed with respect to digital education and information security in university higher education.

Likewise, the opening and development of new lines of research related to digital education in Peru is required, whose need becomes more noticeable day by day in a country where, as a result of the COVID-19 pandemic, the traditional educational approach based only on face-to-face education has been drastically changed.

A separate comment deserves the implementation of an intensive training program for university teachers on issues related to digital education and information security by the authorities of the Ministry of Education of Peru, which could be extended to other educational levels.

7 Financing

The research has been fully self-financed.

8 Conflict of Interest

None.

Appendix

Table 5. Questionnaire applied to obstetric students.

Section 1: *Demographic characteristics*
Q1. Age:
(a) 18–29 years
(b) ≥ 30 years
Q2. Sex:
(a) Woman
(b) Man
Q3. Place of residence:
(a) Rural zone
(b) Urban zone

(*continued*)

Table 5. (*continued*)

Section 2: *Yes/no questions about topics of digital education addressed during obstetric students' learning*
Basic principles of digital education
Q1. Basic definitions (Yes or No)
Q2. Relevance of the protection of confidentiality
Q3. Methods of safeguarding confidentiality
Secure keys
Q1. Methods for establishing secure keys
Q2. Use of secure keys
Safe use of social networks and computer applications
Q1. Precautions in the use of social networks
Q2. Precautions in the use of computer applications
Section 3: *Perception of information security of obstetrics students*
Q1. Perception of their information security:
(a) High security
(b) Medium security
(c) Low security

References

1. Abdurazakov, M., Volkova, S., Vasilyeva, P., Matveeva, E., Tyutyunkova, M.: Electronic educational resources as a means of digital education development. In: V 2020 International Scientific Conference on Innovative Approaches to the Application of Digital Technologies in Education, SLET 2020, pp. 188–202. CEUR-WS, Stavropol (2020)
2. Zilberova, I., Alyaeva, M., Shuraeva, F., Petrov, K.: Digital education: cultural aspects of communication in the modern educational environment. In: Rudoy, D., Olshevskaya, A., Ugrekhelidze, N. (eds.) 14th International Scientific and Practical Conference on State and Prospects for the Development of Agribusiness. INTERAGROMASH 2021. E3S Web of Conferences, vol. 273, 12104. EDP Sciences, Les Ulis (2021). https://doi.org/10.1051/e3s conf/202127312104
3. Mukhametzyanov, I.: Digital educational environment, health protecting aspects. J. Sib. Fed. Univ. Humanit. soc. sci. **12**(9), 1670–1681 (2019). https://doi.org/10.17516/1997-1370-0484
4. Larchenko, V., Barynikova, O.: New technologies in education. In: Rudoy, D., Olshevskaya, A., Ugrekhelidze, N. (eds.) 14th International Scientific and Practical Conference on State and Prospects for the Development of Agribusiness. INTERAGROMASH 2021. E3S Web of Conferences, vol. 273, 12145. EDP Sciences, Les Ulis (2021). https://doi.org/10.1051/e3s conf/202127312145.
5. Chen, Y., Shih, W., Lee, C., Wu, P., Tsai, C.: Relationships among undergraduates' problematic information security behavior, compulsive internet use, and mindful awareness in Taiwan. Comput. Educ. **164**, 104131 (2021). https://doi.org/10.1016/j.compedu.2021.104131
6. Wong, B., et al.: Harnessing the digital potential of the next generation of health professionals. Hum Resour Health. **19**(1), 50 (2021). https://doi.org/10.1186/s12960-021-00591-2

7. Rejeb, S., Kouki, N., Dhaoui, A., Jlidi, N., Bellil, K.: A remarkable concept of learning in times of educational confinement social media and pathologist. Tunisle Medicale. **99**(4), 404–409 (2021)

8. Ahlstrom, L., Holmberg, C.: A comparison of three interactive examination designs in active learning classrooms for nursing students. BMC Nurs. **20**, 59 (2021). https://doi.org/10.1186/s12912-021-00575-6

9. Ye, C., Shi, W., Zhang, R.: Research on gray correlation analysis and situation prediction of network information security. EURASIP J. Inf. Secur. **2021**(1), 1–6 (2021). https://doi.org/10.1186/s13635-021-00118-1

10. Dong, S., Nathanial, S., Chavez, M., Li, Y., Escota, G., Stead, W.: Get privacy trending: best practices for the social media educator. Open Forum Infect. Dis. **8**(3), ofab084 (2021). https://doi.org/10.1093/ofid/ofab084

11. Ndlovu, K., Mars, M., Scott, R.: Interoperability frameworks linking mHealth applications to electronic record systems. BMC Health Serv. Res. **21**, 459 (2021). https://doi.org/10.1186/s12913-021-06473-6

12. Elamir, M.M., Al-atabany, W.I., Mabrouk, M.S.: Hybrid image encryption scheme for secure E-health systems. Netw. Model. Anal. Health Inform. Bioinform. **10**(1), 1–8 (2021). https://doi.org/10.1007/s13721-021-00306-6

13. Peek, N., Sujan, M., Scott, P.: Digital health and care in pandemic times: impact of COVID-19. BMJ Health Care Inform. **27**, e100166 (2020). https://doi.org/10.1136/bmjhci-2020-100166

14. Mendoza, W., Carrion, C., Ibarra, M.: New technologies in the process of digital education and the right to education in times of SARS-CoV-2 pandemic at the Universidad Tecnológica de los Andes in southern Peru. In: Callaos, N., Horne, J., Sanchez,,B., Tremante, A. (eds.) 19th Ibero-American Conference on Systems, Cybernetics and Informatics, CISCI 2020, 17th Ibero-American Symposium on Education, Cybernetics and Informatics. SIECI 2020. vol. 2, pp. 148–154. International Institute of Informatics and Systemics, IIIS, Florida (2020)

15. Flores-Cueto, J., Hernández, R., Garay-Argandoña, R.: Information technologies: Internet access and digital divide in Peru. Revista Venezolana de Gerencia. **25**(90), 504–527 (2020)

16. Robinson, L., et al.: Digital inclusion across the Americas and the caribbean. Social Inclusion. **8**(2), 244–259 (2020). https://doi.org/10.17645/si.v8i2.2632

17. Abimbola, S.: Avoiding the road to nowhere: policy insights on scaling up and sustaining digital health. BMJ Glob. Health **4**, e002068 (2019). https://doi.org/10.1111/1758-5899.12909

18. Lopez-Ercilla, I., et .al.: The voice of Mexican small-scale fishers in times of COVID-19: impacts, responses, and digital divide. Marine Policy **131**, 104606 (2021). https://doi.org/10.1016/j.marpol.2021.104606

19. Reggi, L., Gil-García, R.: Addressing territorial digital divides through ICT strategies: are investment decisions consistent with local needs? Gov. Inf. Q. **38**(2), 101562 (2021). https://doi.org/10.1016/j.giq.2020.101562

20. Throuvala, M., Griffiths, M., Rennoldson, M., Kuss, D.: Policy recommendations for preventing problematic internet use in schools: a qualitative study of parental perspectives. Int. J. Environ. Res. Public Health. **18**(9), 4522 (2021). https://doi.org/10.3390/ijerph18094522

21. Xie, Q., Li, K., Tan, X., Han, L., Tang, W., Hu, B.: A secure and privacy-preserving authentication protocol for wireless sensor networks in smart city. EURASIP J. Wirel. Commun. Netw. **2021**(1), 1–17 (2021). https://doi.org/10.1186/s13638-021-02000-7

22. Ouytsel, J.: The prevalence and motivations for password sharing practices and intrusive behaviors among early adolescents' best friendships – a mixed-methods study. Telematics Inform. **63**, 101668 (2021). https://doi.org/10.1016/j.tele.2021.101668

23. Beltran-Aroca, C., Ruiz-Montero, R., Labella, F., Girela-Lopez, E.: The role of undergraduate medical students training in respect for patient confidentiality. BMC Med. Educ. **21**, 273 (2021). https://doi.org/10.1186/s12909-021-02689-6
24. Ranginwala, S., Towbin, A.: Use of social media in radiology education. J. Am. Coll. Radiol. **15**(1), 190–200 (2018). https://doi.org/10.1016/j.jacr.2017.09.010

Tools and Services of the Cloud-Based Systems of Open Science Formation in the Process of Teachers' Training and Professional Development

Maiia V. Marienko[✉] [iD]

Institute of Informationa Technologies and Learning Tools of the NAES of Ukraine,
9 M. Berlynskoho Street, Kyiv 04060, Ukraine

Abstract. The article defines the tools and services of cloud-based systems of open science formation in the process of teachers' training and professional development. The model of the cloud-oriented methodical system of preparation of teachers of natural and mathematical subjects for work in scientific lyceum is substantiated and described. In particular, the use of the tools and services at each level of the methodological system is described. Preliminary stages of designing a cloud-oriented system of training teachers of natural sciences and mathematics to work in a scientific lyceum are considered. The analysis and assessment of the state of use of cloud-oriented systems are carried out. It was found that the classic stages of designing a cloud-oriented system of training teachers of natural sciences and mathematics to work in a scientific lyceum are: analysis of the research problem, formulation of the design goal (tasks, hypotheses, plan), construction of a prototype, test prototype systems), analysis of test results, adjustment of the components of the prototype depending on the analysis and widespread implementation of the designed system.

Keywords: Open science systems · Cloud-based systems · Teaching staff · Teachers · Scientific lyceums · Tools and services of open science

1 Introduction

In recent years, numerous developments around the world have formed a clear and consistent vision of the introduction of the open science paradigm as a driving force for the creation of a new concept of transparent data-driven science. Open science, open access, open data and open source are becoming increasingly popular and necessary. However, widespread implementation of these practices in Ukraine has not yet been achieved. One reason is that researchers aren't sure how sharing materials will affect their careers. At the same time, despite certain risks associated with sharing data, open materials lead to increased citations, media and peer attention, greater opportunities to organize collaborative work on a single research problem, and additional funding. Such findings are evidence that open research is more beneficial to society and the scientific community than traditional closed practice.

S. Wrycza and J. Maślankowski (Eds.): PLAIS EuroSymposium 2021, LNBIP 429, pp. 108–120, 2021.
https://doi.org/10.1007/978-3-030-85893-3_8

1.1 Formulation of the Problem

Given the significant pedagogical potential and novelty of existing approaches to the design of cloud-based open science systems, their formation and use in educational institutions, these issues still require theoretical and experimental research, refinement of approaches, models, methods and techniques, possible ways of implementation. In particular, theoretical and methodological aspects of determining the structure, functions, tools and technologies of designing cloud-based systems of open science in educational institutions, forms and methods of their use in the process of teaching and professional development of teachers remain virtually undeveloped.

With the emergence of new institutions of specialized education (for example, scientific lyceums), the question arises about the readiness of teachers of natural sciences and mathematics to work in such educational institutions. In this regard, there is a problem of changing the content and organization of advanced training and retraining of teachers of science and mathematics. There is a need, in particular, to design a system of teachers' training for work in the scientific lyceum.

The content of teacher training courses is also influenced by the situation in connection with the introduction of quarantine measures in Ukraine related to the spread of COVID-19 (2020–2021), as forms of distance work, blended and distance learning have become widely used. To organize distance and blended learning, teachers need to learn to use technologies and systems that within one platform will ensure the full organization of the learning process without the use of third-party tools. One option is to use cloud services and cloud-based systems.

On January 16, 2020, the Verkhovna Rada of Ukraine adopted the Law "On Secondary Education", according to which grades 10–12 are a profile level, which also requires appropriate specifics in the retraining of subject teachers. The notion of the scientific lyceum is regulated by the special document, that regulates the functioning of this kind of educational establishments. «Scientific lyceum and scientific boarding lyceum conducts educational activities aimed at attracting and preparing young students for scientific and scientific-technical activities [18]».

1.2 Analysis of Recent Research and Publications

M. P. Shyshkina in her work [19] considered several innovative models of cloud-oriented environment formation: the general model of a cloud-oriented learning and research environment, the conceptual model of a cloud-based university environment research component, the modeling of the cloud-based learning and research environment components groups.

In this regard, in the study of S. H. Lytvynova the model of interaction of cloud-based learning environment subjects is substantiated [12]. However, also, S. H. Lytvynova cites two other models: the model of the student's learning environment and the model of the teacher's learning environment. All these models are combined in a cloud-based learning environment.

T. A. Vakaliuk gives a description of the structural model of the cloud-based learning environment for bachelors of computer science training [20]. She believes that the basis of this model should be the process of training a bachelor's of computer science. A group of scientists V. Yu. Bykov, D. Mikulowski, O. Moravcik, S. Svetsky and M. P. Shyshkina substantiated the model of a cloud-oriented open learning and research environment to support joint activities, but previously in [6] the methods of tools selecting and the prospects for their use in higher education systems are described.

Scientific and methodological principles of formation and development of a cloud-based educational and scientific environment in the context of the priorities of open science and the formation of the European Research Area were considered by V. Yu. Bykov and M. P. Shyshkina [5]. Scientists have analyzed not only the methodological principles of environmental design and development (principles of open education, open science) but also specific principles inherent in cloud-based systems.

O. V. Ovcharuk considered in her research modern strategic approaches to the use of digital tools in the educational process and professional activity of teachers that support a new level of communication and interaction of all participants in the educational process, as well as in the development of digital competence of the teacher [17].

Researchers H. M. S. Bakeer and S. S. Abu-Naser studied the architecture of the intelligent learning system, which in turn contains only four components: the domain model, the student (listener) model, the learning module and the user interface (they can be considered as separate models) [3].

The model of the management system of several organizations, which is based on the cloud model of the community is proposed in the work of K. Dubey, M. Y. Shams, S. C. Sharma, A. Alarifi, M. Amoon and A. A. Nasr [9].

The study by F. Bozkurt [4] is devoted to the analysis of the social science teacher training program in terms of skills development in the 21st century. The study showed that the program needs to be improved because teachers had a low level of competence in interaction in the digital environment.

S. Arslan, İ. H. Mirici, H. Öz [2] implemented and evaluated the author's program of advanced training of English teachers in non-formal educational institutions. The two-week online curriculum was designed to meet the professional needs of teachers and tested. The results of the study showed that the program had a significant impact on the knowledge and behavior of teachers. Although most were positive about the program, some teachers suggested conducting a full-time program and extending the duration of the training.

The study by A. U. Kımav and B. Aydın [10] describes the project of a contextual program of teacher training for the use of Web 2.0 tools in EFL lessons. The participants were 122 English teachers who worked at the School of Foreign Languages of Anadolu University. Eight consecutive stages were followed in the curriculum. This project, according to researchers, can be proposed and developed for such institutions that want to increase the competence of their teachers to integrate technology into the educational process.

Allen C., Mehler D. M. [1] emphasize the benefits of introducing open science practices in the early stages of a scientific career. Taking into account the existing experience, it is important to form the ability to choose methods that meet the research question, taking into account the feasibility, to use new tools that facilitate the exchange and documentation of research work effectively and transparently in the process of young scientists training and professional development [1]. The competencies of open science have been considered for the level of PhD, as well as for young scientists. The open science practices have been actively implemented in the practice of research universities, research centers, but should be also introduced in teacher education, which has been studied insufficiently [15].

Scientists have sufficiently considered various models of organization of the educational process using information and communication technologies (ICT). Also, the models of a cloud-based environment, in particular for the training of relevant profiles are proposed. However, the problem of designing a cloud-oriented methodological system for preparing teachers of science and mathematics to work in a scientific lyceum remains insufficiently studied. This is due to the specific features of the educational process in the scientific lyceum and insufficient training of teachers to teach in such kind of general secondary education institutions. The additional rights and obligations provided by the legislation concerning scientific lyceums should also be taken into account. To solve the outlined problem and to build a cloud-oriented environment, it will be necessary to pre-develop its model for its further design and research. The existing models of cloud-oriented environments are not focused on the objectives of preparing teachers of science and mathematics to work in the scientific lyceum so there is a need for further developments in this area. Therefore, the principles already developed or the general structure of construction of similar models can be taken as a basis.

1.3 The Aim of the Study

The study aims to determine the tools and services of forming cloud-oriented systems of open science in the process of teachers' training and professional development.

2 Research Methodology

Research work will be performed based on the provisions of a system approach as a methodological background for the study of pedagogical and social facts, phenomena, processes; provisions of psychological and pedagogical science in the field of use of information and communication technologies in the educational process of educational institutions.

Theoretical general scientific methods were used to solve the tasks set in the work: analysis of psychological and pedagogical, philosophical sources on the research problem to clarify the state of development of the formation and development of cloud-oriented systems of open science, identify research areas, principles and approaches to cloud-oriented open science systems; analysis of current standards and regulations on the use of cloud services in the learning process and informatization of educational institutions; generalization of domestic and foreign experience in the use of cloud services and technologies in higher education institutions to identify development trends, clarify the basic conceptual and terminological apparatus, establish the conceptual foundations of the study; theoretical analysis, system analysis, systematization and generalization of scientific facts and patterns for the development and design of models of cloud-oriented systems, substantiation of the main conclusions and provisions.

The empirical research methods have been used to solve the set tasks: the experimental study of the use of cloud services in institutions of higher pedagogical education of Ukraine, the expert evaluation of survey results, the observation of the initial activity with the use of cloud technologies in educational and scientific activities.

3   The Results of the Study

3.1   The Stages of Designing a Cloud-Oriented System of Training Teachers of Natural and Mathematical Subjects to Work in a Scientific Lyceum

Currently, there is no cloud-based system that would become a tool in the further preparation of teachers of science and mathematics to work in the scientific lyceum. The following universities were selected as experimental sites: Rivne Regional Institute of Postgraduate Pedagogical Education (2019, number of respondents - 45) and Zhytomyr Polytechnic State University (2020, number of respondents - 824) [14]. The analysis of the survey showed that most teachers recognize the need for a teacher of a scientific lyceum in scientific activities. Most respondents do not use English-language resources and services due to their low level of language proficiency. It has been found that one of the most important ways to get involved in science, according to math teachers, is to participate in scientific conferences. Regarding the implementation and use of the obtained research results, teachers plan to choose the following ways: publication of methodological materials and self-implementation. The analysis of the results of the survey of educators showed that among the respondents 574 people (69.7%) prefer to use information and communication technologies on a local computer. Analysis of the results of the ascertaining stage of the pedagogical experiment shows that there is a problem of preparing teachers of natural sciences and mathematics to work in the scientific lyceum, which needs further solution through preliminary testing and implementation of a specially created cloud-based methodological system.

Scientists have already studied the design as a pedagogical problem. It was found that the pedagogical design can be considered in a broad and narrow sense. Besides, the most important approaches are the design as long-term planning (expanded) and the design of the actual learning process [22]. The design of a cloud-oriented system will differ in that the evolution of the formation and development of cloud-oriented systems, the current state of development and use of cloud-oriented systems and the conceptual principles and principles of using cloud-oriented systems in pedagogical education systems should be studied in detail.

M. P. Shyshkina in the study [19] clearly outlines the stage of the pilot design and the stage of widespread implementation. In this case, the stage of pilot design includes experimental verification of the prototype, pilot implementation and determines the effectiveness of methods, which determines the components of the required resources. At the same time, the scientist separately mentions the stages of pilot design [19]: target, structural-functional, resource and effective. The pilot design concludes with an analysis of the results to adjust the individual components of the cloud-based system (prototype) and the broad implementation phase begins. This stage is completely based on the results obtained at the stage of the pilot design and takes into account all the identified patterns and characteristics of the cloud-oriented system. These properties are further generalized during widespread implementation.

S. P. Lytvynova in his study [13] gives seven main stages of design: problem-educational; content-target; conceptual; component-evaluation; design and modeling; experimental and corrective; evaluative and generalizing. At the same time, the scientist preliminarily analyzes the very concept of didactic design and notes that currently there is no single list of design stages. Each scientist, depending on the design goal determines its number of stages and clarifies their content.

The procedure for designing a cloud-based learning environment for bachelors of computer science is presented by T. A. Vakaliuk [21] consists of seven stages: analysis, problem formulation, formulation of requirements for the cloud-based learning environment, modeling of cloud-based learning environment, development of cloud-based learning environment. educational process, efficiency check, implementation.

If we summarize all the research, we can say that the classic stages of designing a cloud-based system of training teachers of science and mathematics subjects in the scientific lyceum are: the analysis of the research problem, the formulation of design goals (tasks, hypotheses, plan), prototype construction, the testing prototype (cloud-oriented system), the analysis of test results, the adjustment of the components of the prototype depending on the analysis and widespread implementation of the designed system.

3.2 The Model of the Cloud-Oriented Methodical System of Preparation of Teachers of Natural and Mathematical Subjects for Work in Scientific Lyceum

Due to the introduction of quarantine measures caused by the spread of COVID-19 in 2020–2021, distance learning was introduced in most schools of Ukraine. The organization of distance learning is possible through the use of tools of cloud-based systems of open science. Despite the active use of cloud-based systems by educators, there are some problems in the organization of teaching and professional development of teachers. One of the main problems is the lack of methods for using cloud services, which are not localized, but free for use in research and educational activities (cloud services and cloud-based open science systems). There is an assumption that the use of cloud-based systems of open science will make the educational process more scientific and academic, will lead to the solution of certain problems of academic integrity among teachers and students. Therefore, it can be argued that COVID-19 directly affects the purpose of the methodical system of preparation of teachers of natural and mathematical subjects for work in a scientific lyceum (Fig. 1).

The block methodical system of preparation of teachers of natural and mathematical subjects for work in a scientific lyceum unites three levels of implementation, each of which is a separate technique (Fig. 1). The block methodical system of preparation of teachers of natural and mathematical subjects for work in a scientific lyceum is key. This unit, more precisely, its formation is influenced by the goal: to improve the preparation of teachers to work in the scientific lyceum. This goal was formed based on three components: approval of the regulations on the scientific lyceum, the introduction of quarantine measures in Ukraine and the direction of education.

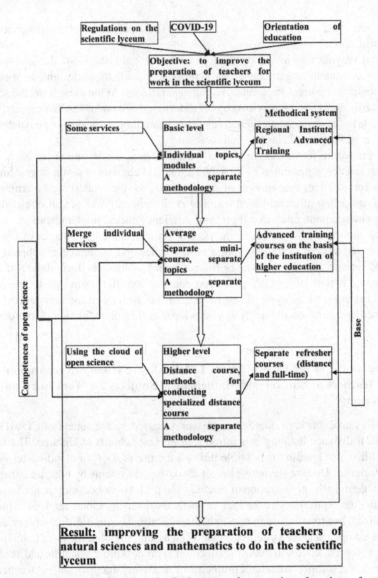

Fig. 1. Model of the cloud-oriented methodical system of preparation of teachers of natural and mathematical subjects for work in the scientific lyceum.

The block of the methodical system of preparation of teachers of natural and mathematical subjects for work in a scientific lyceum will be considered separately. However, its components include three main blocks, the introduction of a methodological system of training teachers of natural sciences and mathematics to work in the scientific lyceum at: basic, intermediate, and advanced levels. These levels are characterized by separate techniques that ensure the formation of competencies of open science.

These methods depend on the meringue, the experimental sites on which the implementation will be carried out: based on regional institutes of advanced training, advanced training courses based on free economic zones and separate advanced training courses (distance and full-time). Each level requires the involvement of certain cloud services of open science. Thus, at the basic level, the use of separate cloud services of open science, located on different platforms. The middle level already involves the use of a group of similar cloud services. At the highest level of the methodical system of training teachers of natural sciences and mathematics to work in the scientific lyceum, the tools of the open science cloud and combinations of its different services are used directly (but based on one platform - the European open science cloud).

Thus, at the basic level, testing takes place within some topics or modules and involves the study of at least one cloud service of open science. The middle level involves not only a survey of existing cloud services but at least their groups, simultaneously used for different activities. Approbation takes place by teachers studying a separate mini-course or individual topics. The highest level ensures the study and use of the tools of the European cloud of open science in a distance course or a specialized distance course. The method for studying these courses is given, as there are certain features in mastering the material at the highest level.

As a result, we will improve the preparation of teachers of natural sciences and mathematics to do in the scientific lyceum.

3.3 Components of the Methodical System of Preparation of Teachers of natural and Mathematical Subjects for Work in a Scientific Lyceum

The purpose of creating a cloud-oriented methodological system for training teachers of science and mathematics to work in the scientific lyceum is to reform teacher training, improve its quality, accessibility and competitiveness. In particular, the cloud-based methodological system is focused on overcoming certain problems identified in the National Strategy for Education Development in Ukraine until 2021, for example: "unpreparedness of some educators for innovation" [16].

The methodical system of training teachers of natural and mathematical subjects to work in the scientific lyceum will be several separate methods of using cloud services (or cloud-oriented systems). The methodological system includes: methods of using cloud services that provide search, collection, accumulation of data; methods of using cloud services for presentation, processing, visualization of patterns in the data; methods of analysis and interpretation of the results using cloud services; methods for validation, discussion, collective evaluation of conclusions, review within the cloud-based system and further implementation and publication of the results.

The purpose for the creation of a methodical system of training is the formation of a cloud-based system of training teachers of natural sciences and mathematics to work in the scientific lyceum. An open science platform (with separate tools) will act as a cloud-oriented system, in particular, the use of its components in the educational process is envisaged. This will be facilitated by wider access to the tools of the European cloud of open science and increase the level of scientific organization of education in the scientific lyceum.

The content of the methodical system of training teachers of natural sciences and mathematics to work in the scientific lyceum is aimed at forming ICT competence of teachers and students to use cloud services and cloud-based systems at each stage of research and in the educational process.

Teaching methods used in the cloud-based methodological system: verbal (video lectures, text chats, online conversations); visual (video instruction, training, seminar-training); practical (practical work, group performance of tasks). Forms of study: lectures; practical work; group work; individual work; training sessions; work in research network projects; explanations and individual consultations, control check. The main emphasis is on the organization of work in groups, as this form of training is key to the organization of project activities. Teaching aids: tools of the European cloud of open science (some cloud services and systems that can be used in the educational process); cloud services that are not part of the European Open Science Cloud, but whose use does not contradict the principles of open science.

The tools and services of forming cloud-oriented systems of open science in the process of teachers' training and professional development are the important component of the proposed methodical system. In the course of the study, the special kinds of tools and services were considered and used as appropriate for the purpose of teachers' training.

The teacher profiles in open access: Google Scholar, ORCID, Web of Science ResearcherID, Scopus, and Bibliometrics of Ukrainian science.

The services for finding scientific works: Google Scholar, arXiv.org, Electronic Library of the National Academy of Pedagogical Sciences of Ukraine, and the dblp computer science bibliography.

The services for the publication of scientific achievements of teachers: Electronic Library of the National Academy of Pedagogical Sciences of Ukraine, arXiv.org, e-journal "Information Technologies and Learning Tools".

The services of the European Open Science Cloud, in particular, the electronic learning resources concerning the learning subject domains.

These kinds of open science tools and services were used to support the main stages of scientific research considered in the course of the research.

1. Search, collection, accumulation of data on the problem of research and its coverage in the literature, ascertaining data. The most common cloud services: Google Academy, electronic libraries of institutions, repositories, archives of open access materials, international databases, scientometric databases.
2. Presentation, processing, visualization of patterns in the data, including sharing. The most common cloud services: spreadsheets (Microsoft Office Excel), Google spreadsheets, cloud systems of computer mathematics (SCM).
3. Analysis and interpretation of the results (for example, using statistical packages. The most common cloud services: statistical packages, presentation, or publication services.
4. Validation, discussion, collective evaluation of conclusions, peer review. The most common cloud services: social networks, cloud system tools, virtual interactive whiteboards.

5. Implementation, publication, use. The most common cloud services: personal sites, blogs, social networks, educational portals.

The result component: expanding access to cloud services and cloud-based systems, increasing the level of organization of research in scientific lyceums, increasing the level of ICT competence of teachers and students.

4 Analysis and Evaluation of Possible Ways for Future Developments and Research Prospects

Since at the highest level model of the cloud-oriented methodical system of preparation of teachers of natural and mathematical subjects for work in the scientific lyceum provides use of separate cloud services of the European open science cloud it is necessary to develop its technical possibilities and communication with edge computing.

The increasing diversity of architectural approaches we face inevitably affects the development of cloud-based systems (including open science). Whereas previously data were processed and made available on a centralized server, it now includes only one of several architectural approaches. Depending on the specific use case, calculations may also take place in [11]:

1. boundary calculations (for example, on touch devices);
2. foggy (for example, on network gateways);
3. directly in the cloud.

New computing technologies have led to the emergence of data spaces of varying complexity, which also occur at different levels, such as local or regional. If most data are processed in the cloud today (80% according to a recent study by the International Data Corporation), it is expected that the rapid spread and use of the Internet of Things (IoT) will lead to most computing at the edge of the network. Today, network latency creates limitations for the use of edge and fog computing, which will be addressed by fifth-generation wireless technology for digital cellular networks (5G). Despite the different architectural structures of boundary, fog, and cloud computing, a common feature of the three approaches is the combination of data and algorithms.

Storage capacity has been steadily increasing since the first hard drives (more than 60 years ago). Modern technology currently limits the storage capacity of hard drives (approximately 10 TB per device), and technology to support 100 TB hard drives is expected to appear by 2025 [7]. However, many modern applications deal with data sets that are larger than what is appropriate for a single machine or server. Moreover, these programs potentially require varying degrees of performance, availability, and resiliency. Programs are becoming more distributed in-network architecture, collecting data from several different locations, such as in touch devices, or by allocating computing to different locations, as in applications for edge computing. To simplify the process of accessing and processing data, you should use a distributed data management system. Distributed Data Management System (DDMS) is integrated into different management systems managed by different domains, to create a single view of user data across administrative, geographical and technological boundaries. This is the principle of operation of boundary calculations.

A study [23] dating back to 2020 highlighted the integration of EOSC with computational scenarios of foggy technologies, edge technologies and the Internet of Things as a problem, as some IPs have extensive sensor networks and technologies that need to be connected to a wider electronic infrastructure. That is, as of 2020, the European cloud of open science did not use the technology of boundary computing.

Further development and deployment of modern technologies, such as cloud and edge computing, big data, AI, IoT and Digital Twins, will enable the full virtualization of the pan-European RI platform as Services (PRIaaS), enabling virtualized RI on-demand provisioning for specific scientific domains and communities; The latest data management and data processing technologies will allow the full implementation of FAIR principles and reliable data exchange, supporting the entire data life cycle and value chain with the necessary infrastructure services. It is expected that the implementation of 5G technologies will begin the preparatory phase of the EOSC-I phase in individual projects and test samples and will become the main technology that will contribute to network virtualization and RI in the future, combined with Virtual Private Cloud (VPC) [8] technologies supported by modern cloud platforms.

The processes of data processing are the important stage of any research work that's why the services of European open science cloud are the necessary kind of tools and services of the methodical system of teachers' training. There are new ways of using these tools in the learning process that need methodological investigation and testing. Some kinds of these services were included in the program of teachers training developed during the research project "Cloud-oriented systems of open science in teaching and professional development of teachers" (2020–2022).

5 Conclusions and Discussion

According to the study, the stages of designing a cloud-based system of training teachers of natural sciences and mathematics to work in a scientific lyceum are: the analysis of the research problem, the formulation of the design goal, the construction of a prototype, the testing of a prototype, the analysis of test results, the adjusting test components depending on the analysis and widespread implementation of the designed system.

The study presents a description of the model of the cloud-oriented methodological system of training teachers of natural sciences and mathematics to work in a scientific lyceum. This model contains three main levels of testing: basic, intermediate, and advanced. Each level has its methodology, which involves the use of special cloud services, their groups, or directly the tools of the European open science cloud.

The main components of the methodical system of preparation of teachers of natural and mathematical subjects for work in the scientific lyceum were considered in the research. The purpose, content, methods, forms and tools of teaching are described. As of 2020, the European Open Science Cloud did not use edge computing technology. It has been found that this problem exists and research is underway to address the integration of the EOSC with computational scenarios of fog technologies, edge technologies and the Internet of Things.

The main stages of scientific research are considered and the use of cloud services of open science at each stage of scientific research is offered. Selection of cloud services is performed. Considered: joint data processing services, joint project work services, and video conferencing services as joint work services. The given classifications will be useful for the further organization of each stage of scientific research.

A detailed description of the methods of using the services of the cloud-oriented system of training teachers of natural and mathematical subjects to work in the scientific lyceum will serve as a prospect for further explorations.

Acknowledgments. The article presents the results obtained during the implementation of the project of the National Research Fund of Ukraine "Cloud-oriented systems of open science in teaching and professional development of teachers" (DR No. 0120U104849) in 2021, the responsible performer of which was the author of the article.

References

1. Allen, C., Mehler, D.M.: Open science challenges, benefits and tips in early career and beyond. PLoS Biol. **17**(5), e3000246 (2019). https://doi.org/10.1371/journal.pbio.3000246
2. Arslan, S., Mirici, İ.H., Öz, H.: Implementation and evaluation of an EFL teacher training program for non-formal education settings. Elem. Educ. Online **3**, 1337–1370 (2020). https://doi.org/10.17051/ilkonline.2020.729666
3. Bakeer, H., Abu-Naser, S.: An intelligent tutoring system for learning TOEFL. Int. J. Acad. Dev. **2**, 9–15 (2019)
4. Bozkurt, F.: Evaluation of social studies teacher training program in terms of 21st century skills. Pamukkale Univ. J. Educ. **51**, 34–64 (2021). https://doi.org/10.9779/pauefd.688622
5. Bykov, V., Shyshkina, M.P.: The conceptual basis of the university cloud-based learning and research environment formation and development in view of the open science priorities. Inf. Technol. Learn. Tools **68**, 1–19 (2018). https://doi.org/10.33407/itlt.v68i6.2609
6. Bykov, V., Mikulowski, D., Moravcik, O., Svetsky, S., Shyshkina, M.: The use of the cloud-based open learning and research platform for collaboration in virtual teams. Inf. Technol. Learn. Tools **76**, 304–320 (2020). https://doi.org/10.33407/itlt.v76i2.3706
7. Cushing, R., et al.: Process data infrastructure and data services. Comput. Inform. **39**, 724–756 (2020). https://doi.org/10.31577/cai_2020_4_724
8. Demchenko, Y., de Laat, C., Los, W.: Future scientific data infrastructure: towards platform research infrastructure as a service (PRIAAS). In: The 2020 International Conference on High Performance Computing Simulation (HPCS 2020), The 18th Annual Meeting, Barcelona, Spain (2021). http://hpcs2020.cisedu.info/virtual/Onlineevent
9. Dubey, K., Shams, M.Y., Sharma, S.C., Alarifi, A., Amoon, M., Nasr, A.A.: A management system for servicing multi-organizations on community cloud model in secure cloud environment. IEEE Access. **7**, 159535–159546 (2019). https://doi.org/10.1109/ACCESS.2019.2950110
10. Kımav, A.U., Aydın, B.: A blueprint for in-service teacher training program in technology integration. J. Educ. Technol. Online Learn. **3**(3), 224–244 (2020)
11. Kotsev, A., Minghini, M., Tomas, R., Cetl, V., Lutz, M.: From spatial data infrastructures to data spaces-a technological perspective on the evolution of european sdis. Int. J. Geo-Inf. **9** (2020). https://doi.org/10.3390/ijgi9030176
12. Lytvynova, S.H.: Design of cloud-oriented educational environment of a comprehensive educational institution. CK "Komprint", Kyiv (2016)

13. Lytvynova, S.H.: Theoretical and methodological foundations of designing a cloud-based learning environment of a secondary school, dis. Dr. Ped. Sciences, Ph.D. thesis, 13.00.10. Nat. acad. ped. of Sciences of Ukraine, Institute of Information Technologies and Learning Tools, Kyiv (2016)

14. Marienko, M.: Analysis of the state of the problem of preparation of teachers of natural and mathematical subjects for work in the scientific lyceum. J. Inf. Technol. Educ. (ITE) **43** (2020). https://doi.org/10.14308/ite000719

15. O'Carroll, C., et al.: Written by The Working Group on Education and Skills under Open Science Providing researchers with the skills and competencies they need to practise Open Science Report of the Working Group on Education and Skills under Open Science (2017). https://doi.org/10.2777/121253

16. On the national strategy for the development of education in Ukraine until 2021. https://zakon.rada.gov.ua/laws/show/344/2013#Text

17. Ovcharuk, O.V.: Current approaches to the development of digital competence of human and digital citizenship in European countries. Inf. Technol. Learn Tools **76**, 1–13 (2020). https://doi.org/10.33407/itlt.v76i2.3526

18. Resolution on approval of the Regulations on scientific lyceum and scientific boarding lyceum. https://zakon.rada.gov.ua/laws/show/438-2019-%D0%BF?lang=uk#Text

19. Shyshkina, M.P.: Formation and development of the cloud-based learning and research environment of higher education institution. UkrISTEI, Kyiv (2015)

20. Vakaliuk, T.A.: Theoretical and methodical principles of the cloud-based learning environment design and use in the training of bachelors in computer science. Ph.D. thesis, 13.00.10. Nat. acad. ped. of Sciences of Ukraine, Institute of Information Technologies and Learning Tools, Kyiv (2019)

21. Vakaliuk, T.: Model of cloud-oriented system of support for teaching bachelors of computer science. Inf. Technol. Learn. Tools **56**, 64 (2016). https://doi.org/10.33407/itlt.v56i6.1415

22. Yaroshynska, O.O.: Designing the educational environment of professional training of future primary school teachers as a pedagogical problem. Probl. Modern Teacher Train. **10**, 110–119 (2014)

23. Zhao, Z., Jeffery, K., Stocker, M., Atkinson, M., Petzold, A.: Towards operational research infrastructures with FAIR data and services. In: Zhao, Z., Hellström, M. (eds.) Towards Interoperable Research Infrastructures for Environmental and Earth Sciences. LNCS, vol. 12003, pp. 360–372. Springer, Cham (2020). https://doi.org/10.1007/978-3-030-52829-4_20

Innovative Methods in Data and Process Analysis

Convolutional Neural Networks to Protect Against Spoofing Attacks on Biometric Face Authentication

Alexandr Kuznetsov[1,2]($\boxtimes$) (ID), Sergey Fedotov[1] (ID), and Mykhaylo Bagmut[1] (ID)

[1] V. N. Karazin Kharkiv National University, Svobody sq., 4, Kharkiv 61022, Ukraine
kuznetsov@karazin.ua
[2] JSC "Institute of Information Technologies", Bakulin St., 12, Kharkiv 61166, Ukraine

Abstract. Modern technologies of authentication and authorization of access play a significant role in ensuring the protection of information in various practical applications. We consider the most convenient and used in modern mobile gadgets face authentication, ie when the primary information to provide access are certain features of biometric images of the user's face. Most of the systems use intelligent processing of biometric images, in particular, artificial intelligence technology and deep learning. But at the same time, as always in cybersecurity, technologies for violating biometric authentication are being studied and researched. In particular, to date, the most common attack is substitution (spoofing), ie when attackers use pre-recorded biometric images to gain unauthorized access to critical information. For example, this could be a photo and/or video image of a person used to unlock their smartphone. Protection against such attacks is very difficult, because it involves the development and study of technologies for detecting signs of life. The most promising in this direction are artificial intelligence techniques, in particular, convolutional neural networks (CNN). This is the practical application of intelligent processing of biometric images and is studied in this article. We review various CNN settings and configurations and experimentally investigate their effect on the effectiveness of signs of life detection. For this purpose, success and failure indicators of the first and second kind are used, which are estimated by the values of cross entropy. These are reliable and reproducible indicators that characterize the effectiveness of protection against spoofing attacks on biometric authentication on the face. The world-famous TensorFlow and OpenCV libraries are used for field experiments, photos and videos of various users are used as source data, including Replay-Attack Database from Idiap Research Institute.

Keywords: Authentication technologies · Convolutional neural networks · Spoofing attacks · Biometric images · Face authentication · Cybersecurity

1 Introduction

Biometric authentication is a process of providing protection, which is based on the unique biological characteristics of a person to confirm the identity of the user [1, 2].

S. Wrycza and J. Maślankowski (Eds.): PLAIS EuroSymposium 2021, LNBIP 429, pp. 123–146, 2021.
https://doi.org/10.1007/978-3-030-85893-3_9

Modern biometric authentication systems compare incoming biometric data with stored validated authentic data in the database [2, 3]. If both biometric data samples match, then the authentication is confirmed.

Typically, biometric authentication is used to control access to physical and digital resources, such as buildings, rooms, and computing devices. [4–6].

Thus, modern biometric authentication systems use the unique physiological characteristics of humans. The most common are such systems:

- Fingerprint authentication (approximate probability of false positives (FP) – 0.001%) [7–10];
- Authentication by the iris (FP is almost absent) [11–13];
- Authentication by palm geometry [14, 15];
- Voice identification [16–18];
- Facial authentication (FP is almost absent) [18–20], and others [21–24].

Despite the relatively high levels of protection against FP, modern biometric authentication systems are often at risk of so-called "spoofing" attacks, or substitution attacks [25–27]. The essence of such an attack is to create an object that exactly repeats the features of the biometric key [28, 29]. For example, in fingerprint authentication systems, this may be a rubber copy of the scanned finger [10], or a photograph of a person in facial authentication systems [30].

It can be quite difficult to defend against such attacks, because by accurately repeating all the characteristics of the biometric key, an attacker can deceive detectors and gain unauthorized access. Therefore, it is necessary to use new intelligent technologies for recognition of biometric patterns, which, in addition to extracting control attributes, can also guarantee the integrity of biometrics, i.e. allow you to detect signs of life. Our motivation is to research this particular area of intelligent technologies for detecting life in biometric images, which will help protect against advanced spoofing attacks.

This article investigates artificial intelligence technologies to protect against spoofing attacks on biometric facial authentication systems. In particular, convolutional neural networks [31, 32], which are most often used for image processing and to detect signs of vitality of the human face, are considered. Various network settings and configurations that affect the effectiveness of the detection of signs of vitality are investigated. According to the results of research, empirical dependences are obtained that characterize the effectiveness of protection against spoofing attacks on biometric authentication on the face.

2 Related Works

A large number of scientific works are devoted to the study of neural networks. In particular, even in the first articles [33, 34] the theoretical basis was laid and the huge potential of artificial intelligence was noted. Today it is a separate branch of scientific research that is rapidly developing and expanding into various spheres of human life [31, 35, 36].

One of the most interesting practical applications of neural networks is to solve complex problems of biometric authentication. Convolutional neural networks proved to be the most suitable for this purpose [32, 35]. For example, [7, 19] uses neural network technologies to increase the efficiency and security of fingerprint authentication. In Articles [11, 12], neural networks are used to process biometric data of the iris. In work [37] implicit authentication of users on the basis of research of features of their course is offered. A specially trained artificial neural network was also used for this purpose. In [21] neural networks were used for authentication by the image of the veins of human hands, and in [20] by the image of the human face. The latter direction is the most relevant, because a person's face is easy to scan with a smartphone camera, i.e. such studies, in our opinion, are the most suitable for practical implementation [3, 38].

It should be noted that in parallel with the research of secure authentication, possible ways of security breach are usually also studied. In particular, in recent years there has been a tendency to spread the so-called spoofing attacks, when attackers try to disrupt the authentication system by substituting real biometric images with pre-recorded copies [39–41].

Various techniques are used to detect spoofing attacks. For example, in [41, 42] possible distortions of biometric images are investigated. In possible distortions of biometric images are investigated. In [43] possible anomalies are investigated, which are classified and stored in appropriate databases such as signature intrusion detection systems. In [44] the technique of analysis of visual artifacts collected in the so-called visual code books and this, in the opinion of the authors, leads to increased effectiveness in counteracting spoofing attacks. In [45] the use of Bloom filters was studied, and in [46] multispectral visualizations as ways to counteract biometric attacks. However, the most effective is the detection of counterfeit attacks using artificial intelligence methods [47, 48]. Indeed, the detection of forgery of biometric images can be presented as a task of detecting signs of life [49–51]. Artificial neural networks provide a powerful mechanism for detecting signs of life [30, 52], and therefore can be used to counter spoofing attacks [38].

It should be noted that the above results from related works make mainly a theoretical contribution to the development of technologies for detecting spoofing attacks. It is important to practically test various technologies for detecting life, to study the possibilities of their application to detect spoofing attacks, as well as to quantify successful detection, for example, in the form of errors of the first and second kind. This is precisely the contribution of our work. In our research, we used convolutional neural networks (CNN) built using the TensorFlow library and the OpenCV library [53]. A Replay-Attack Database from the Idiap Research Institute was also used to train the artificial network and to conduct testing [54]. The aim of our study is to experimentally test the possibility of using CNN to detect signs of life, justify the settings and parameters of an artificial neural network to protect against spoofing attacks on biometric face authentication.

3 Neural Network Setup and Configuration

A combined image analysis approach, namely texture and motion analysis, was used to set up the neural network. This choice is justified by the simplicity of the implementation of the algorithm, which gives a fairly low rate of false positives.

Work on the neural network can be divided into four logical stages:

- Creating a set of initial data with originals and forgeries of biometric keys;

- Implementation of feature selection algorithms using convolutional neural networks;
- Training of the model on the basis of the received initial data;
- Network testing and refinement based on work results.

3.1   Creating a Set of Initial Data

Two videos are used as the initial data. The first shows a man moving around the room, looking at the camera. This video is the original and serves as an example of a real person. The second video contains the first, but removed from the smartphone screen and acts as a video forgery.

The following steps are performed to form training kits:

- The video is split into images at intervals;
- With the help of the face detector Caffe (built into the OpenCV library), the ROI (Region of Interest - rectangular areas of interest) of faces from the photo is selected and stored. Using Caffe allows you to focus directly on recognizing spoofing attacks and not waste time on implementing a face detector and creating data sets for it;
- The stored ROIs form two data sets of original and fake faces, which are then used to train the neural network.

This set of data allows you to create an active neural network, but with quite important limitations - since there is only one video sequence with one person, the neural network will be really active for people with similar physiological properties. Therefore, to improve the results of the network, you can create several videos, which will be people of different ages, genders and ethnicities.

Also equally important are the camera's resolution and shooting location. You can improve the network result by shooting people on different cameras and in different light levels.

After such steps, the data set will be as complete as possible, and the results of the network - as accurate as possible under any circumstances.

3.2   Implementation of Feature Selection Algorithms

The capabilities of the TensorFlow and OpenCV libraries were used to implement feature selection algorithms.

TensorFlow is an open platform for machine learning. It has a flexible system of tools, libraries and resources, just to create models of convolutional neural networks. Keras high-level application programming interface is used for such purposes.

Keras is an open, high-performance interface for solving machine learning problems with an emphasis on modern deep learning. It provides basic abstractions and building blocks for developing and transporting machine learning solutions with high iteration rates. Keras integrates deeply with TensorFlow's low-level functionality, allowing you to develop workflows where any part of the functionality can be customized.

OpenCV (Open Source Computer Vision Library) is a library of programming functions, mainly aimed at real-time computer vision. Originally developed by Intel, it was later supported by Willow Garage, and then by Itseez (which was later acquired by Intel).

The entire network can fit into one static method, which receives a face image as input and returns a neural network model. The parameters of this method will be the size of the input image, the number of channels (3 in our case, because we use raster RGB images) and the dimension of the classification (2 in our case, because we have only two classes – "real" and "fake" face).

One of the important steps in the development of a neural network is the choice of the transfer function, or activation function. In the General case, in artificial neural networks, the transmitting function of a node determines the output of this node based on input data or a set of input data. In other words, it forms some dependence of the output signal of the neuron on the input.

The ReLu (rectified linear unit) function was chosen as the activation function for the network operation, which is quite often chosen for convoluted neural networks due to the speed of action and learning. The description of the function is shown in Fig. 1.

This transfer function was introduced for dynamic networks by Hahnloser and others in 2000 with a biological basis and a mathematical rationale. ReLU is, as of 2018, the most popular transmission function for deep neural networks.

ReLU (rectified linear unit)

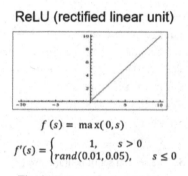

$$f(s) = \max(0, s)$$

$$f'(s) = \begin{cases} 1, & s > 0 \\ rand(0.01, 0.05), & s \leq 0 \end{cases}$$

Fig. 1. ReLu activation function

Using the Keras API, we will build the structure of a convolutional neural network. Network implementation should work with a small amount of resources, so the number of network layers will be small.

All layers can be divided into two sets, which will be identical in structure, but will differ in parameters. The general appearance of one set of layers will look like:

$$Convolution - ReLu - Convolution - ReLu - Merge.$$

After each such set, batch normalization and dropout operations are also performed. Batch normalization applies a transformation that maintains an average output value of about 0 and a standard deviation of the output of about 1. Dropout is a technique used to prevent the model from oversaturating. Dropout works by randomly setting the source edges of hidden neurons (neurons that make up hidden layers) to 0 at each update of the learning phase.

After passing such a set of layers and operations, we get a map of the selected features. Passing the second set will double the size of the map. Therefore, the parameters of the

dimension of the convolution layers of the first and second set will differ twice – 16 ×
16 in the first (Fig. 2) and 32 × 32 in the second (Fig. 3).

After passing two sets of layers the received result arrives on final operations (Fig. 4),
namely smoothing, consolidation, activation and normalization with dropout.

```
model.add(Conv2D(16, (3, 3), padding="same",
    input_shape=inputShape))
model.add(Activation("relu"))
model.add(BatchNormalization(axis=chanDim))
model.add(Conv2D(16, (3, 3), padding="same"))
model.add(Activation("relu"))
model.add(BatchNormalization(axis=chanDim))
model.add(MaxPooling2D(pool_size=(2, 2)))
model.add(Dropout(0.25))
```

Fig. 2. The first set of neural network layers

```
model.add(Conv2D(32, (3, 3), padding="same"))
model.add(Activation("relu"))
model.add(BatchNormalization(axis=chanDim))
model.add(Conv2D(32, (3, 3), padding="same"))
model.add(Activation("relu"))
model.add(BatchNormalization(axis=chanDim))
model.add(MaxPooling2D(pool_size=(2, 2)))
model.add(Dropout(0.25))
```

Fig. 3. The second set of neural network layers

```
model.add(Flatten())
model.add(Dense(64))
model.add(Activation("relu"))
model.add(BatchNormalization())
model.add(Dropout(0.5))
```

Fig. 4. Forming a network response layer

The role of the smoothing layer is extremely simple: the smoothing operation on the
tensor transforms the tensor to have a shape equal to the number of elements contained
in the tensor, not including the batch size.

The compaction layer is the only actual network layer in this model. The compact
layer supplies all the output of the previous layer to all its neurons, each neuron provides
one output to the next layer.

The compaction layer is the only actual network layer in this model. The dense layer
supplies all the output of the previous layer to all its neurons, each neuron provides one
output to the next layer.

At the end, the model goes through another compaction layer with a parameter equal to the classification dimension (in our case 2) and the Softmax activation function - a generalization of the logistic function that "compresses" the K-dimensional vector z with arbitrary component values to the K-dimensional vector $\sigma(z)$ with the actual values of the components in the area [0, 1] that add up to one.

At the output we get a model of a convolutional neural network. The resulting model can be taught using the initial data created in the first stage.

3.3 Network Training and Testing

To begin with, training parameters are created. We store the following data in variables:

- INIT_LR – initial accuracy of training;
- BS – the size of the training unit;
- EPOCHS – number of learning iterations.

The training process begins with the processing of initial data. Images are read, divided into groups of real and fake with a pointer in the file name. All images are scaled to a size of 32×32 and entered into the data structure - a list.

The list is then converted to an array of the NumPy library and the pixel intensity values are converted to values from the range 0..1 by dividing them by 255. The resulting structure is divided by a ratio of 3 to 1, 1 part will be used for testing and 3 - for learning.

Additionally, a generator of random augmentations, ie distortions of input images, is created. It will distort the input data by scaling, rotating, mirroring, and so on. Such distortions will increase the learning efficiency of the convolutional neural network and significantly increase the probability of its correct operation with different input data.

A model of a convolutional neural network is built, according to the principle described above. Formed study groups are presented to the model in the form of pairs "image - answer". Training takes place using the fit API Keras method (Fig. 5). Its functionality is to train the neural network model to adjust all parameters to the correct value for a correct comparison of input and output data.

```
print("[INFO] training network for {} epochs...".format(EPOCHS))
H = model.fit(x=aug.flow(trainX, trainY, batch_size=BS),
    validation_data=(testX, testY), steps_per_epoch=len(trainX) // BS,
    epochs=EPOCHS)
```

Fig. 5. Neural network learning script

The learning results of the convolutional neural network can be tracked by graphically displaying the values of erroneous and correct operations of the algorithm during a given number of iterations.

Using the tools of the matplotlib library, we will construct a graph of 4 curves that characterize the effectiveness of network learning, namely the curves of errors and successful operation of learning and testing. The resulting graph is shown in Fig. 6.

The figure shows:

- where train_loss – training errors;
- val_loss – testing errors given in the form of cross-prediction entropy and correct prediction error response;
- train_acc and val_acc – successful training and testing results, respectively.

The error rates for train_loss and val_loss are set using the error functions of the Keras library. The purpose of such functions is to calculate the error rate, which the model should seek to minimize during training.

Thus, in Fig. 6 binary cross entropy is used as a function of errors. It determines the average number of bits needed to recognize an event from the event space, if the coding scheme used is based on the probability distribution q (prediction), instead of the "true" distribution p (correct answer).

Cross entropy is calculated by the formula:

$$H_p(q) = -\frac{1}{N} \sum_{i=1}^{N} y_i * \log(p(y_i)) + (1 - y_i) * \log(1 - p(y_i)),$$

where N – sample size, y is the correct answer ("true" or "fake"), $p(y_i)$ – is the predicted probability that y is "true".

The paper also uses another error function, namely the root mean square prediction error. It is calculated by the formula:

$$MSE = \frac{1}{N} \sum_{i=1}^{N} y_i - p(y_i).$$

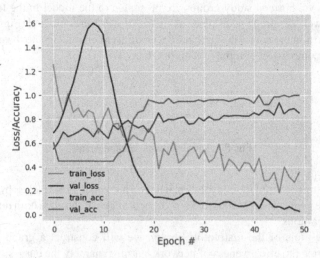

Fig. 6. Neural network training graph with parameters INIT_LR = 1e−4 and BS = 8

4 Experimental Research

The purpose of our experiments is to study convolutional neural networks for detecting signs of life and, ultimately, for detecting spoofing attacks. The methodology of experiments consists in studying various parameters of neural networks, as well as studying their influence on the speed and accuracy of training. In particular, we examine the parameters: initial precision (INIT_LR); training block size (BS); number of training iterations (EPOCHS). We also consider different activation functions: ReLU; sigmoid function; ELU; SELU.

4.1 Research of Network Training Parameters

Consider the behavior of the values of training accuracy for different values of learning parameters, namely the initial learning speed and the size of one game.

Learning speed is a hyperparameter that controls the scale of the model change in response to the calculated error each time the scales are updated. Choosing a learning speed is a difficult task, because too little value can lead to a long learning process, while too much value can lead to the acquisition of a suboptimal set of weights too fast or unstable learning process.

Learning speed can be the most important hyperparameter when setting up a neural network. Therefore, it is vital to know how to investigate the effect of learning speed on model performance and to build an intuition about the dynamics of learning speed on model behavior.

Regarding the neural network created in Sect. 3, by modifying the learning speed parameter, we can observe a change in the learning curve. For example, in Fig. 7 shows a graph with the parameters $INIT\_LR = 1e-1$ and $BS = 8$.

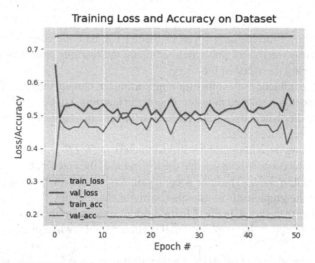

Fig. 7. CNN learning curve with parameters INIT_LR = 1e−1 and BS = 8

Hereinafter, the calculation of the root mean square prediction error is used as a function of errors.

Indeed, the graph shows that due to the high speed of learning there was the assimilation of suboptimal set of weights, and therefore the learning process led to almost nothing - CNN gave the correct answer in ~50% of cases, ie almost at random.

In Fig. 8 shows the inverse case when the value of the learning speed parameter is quite small, namely $INIT\_LR = 1e-10$. We can observe confirmation of the words above about the long learning process. Thus, in the same 50 epochs, the neural network was able to achieve only ~55% accuracy.

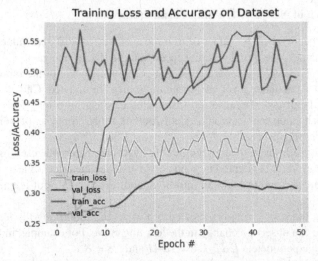

Fig. 8. CNN learning curve with parameters INIT_LR = 1e−10 and BS = 8

The selection of the learning speed parameter can be done empirically, for example, using software analyzers that would compare the learning results of the network, gradually changing the value of the speed.

Consider the influence of another training parameter, namely the size of the batch. In general: larger batch sizes lead to faster progress in learning, but do not always converge so quickly. Smaller batch sizes train more slowly, but can converge faster. It definitely depends on the type of problem.

In Fig. 9 shows a training graph with parameters $INIT\_LR = 1e-4$ and $BS = 1$. By the behavior of the curve we can see that the low value of the batch size led to a slowdown in learning, as noted above. Thus, the accuracy of close to 100% CNN reached only after 25 epochs, but the graph itself is smoother, without sharp jumps.

Compared with the graph in Fig. 6 learning slowed down by about 8 epochs, but starting from the 25th epoch the graph in Fig. 9 is almost stable, while in Fig. 6 fluctuations still continue.

Analyzing the graph in Fig. 10 we see that the training was faster than in the previous case, and the network reached close to 100% accuracy after 10 epochs, but the overall stability of the graph is lower, there are fluctuations, which confirms the above patterns.

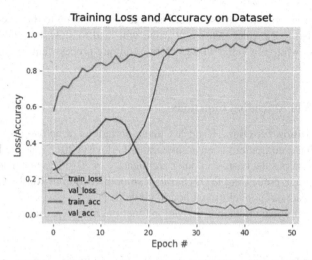

Fig. 9. CNN learning curve with parameters INIT_LR = 1e−4 and BS = 1

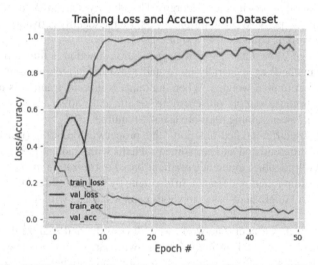

Fig. 10. CNN learning curve with parameters INIT_LR = 1e−4 and BS = 20

Based on the analysis of the influence of neural network learning parameters, we can draw the following conclusions and recommendations for their choice:

- when choosing the initial speed of training, it is necessary to find a balance between the desired duration of the training process and the correct assimilation of the optimal set of weights. Hence, too small a value will lead to slow learning, whereas too large a value will be ineffective and the network will be inoperable;

- when choosing the batch size it is necessary to pay attention to the size of the initial data, and also to consider that small value will lead to slow, but stable and effective training, while big value will accelerate training process, but training efficiency decreases.

4.2 Research of Activation Functions

1. ReLU activation function. There are many variations of activation functions. For example, the basic computer vision architecture of AlexNet 2012 uses the ReLU activation function, as does the basic computer architecture of 2015 ResNet. The main model of BERT language processing in 2018 uses a smooth version of ReLU, GELU. ReLU was used to create our convolutional neural network.

In addition to empirical indicators, activation functions also have different mathematical properties:

- Nonlinearity - when the activation function is nonlinear, then a two-layer neural network can be proved as a universal approximator of functions. This is known as the general approximation theorem. The identity activation function does not satisfy this property. When multiple layers use the identity activation function, the entire network is equivalent to a single-layer model.
- Value Range - when the range of the activation function is finite, gradient-based learning methods are generally more stable, as sample presentations only significantly affect limited weights. When the range is infinite, training is usually more effective because sample presentations significantly affect most weights. In the latter case, lower training rates are usually required.
- Permanently differentiated function - this property is desirable (ReLU is not permanently differentiated and has some problems with gradient-based optimization, but it is still possible) to include gradient-based optimization methods. The binary step activation function does not differentiate by 0, and it differentiates to 0 for all other values, so gradient-based methods cannot succeed with it.

Based on the developed neural network, we will consider and analyze the influence of activation functions on the learning curve. We will start from Fig. 11, which shows a learning curve using the ReLU activation function and parameter values INIT_LR = 1e−4 and BS = 20.

2. Sigmoidal activation function. A sigmoid is a continuously differentiated monotonic nonlinear S-shaped function that is often used to smooth the values of a quantity. Often the sigmoid is understood as the logistic curve shown in Fig. 12, which is determined by the formula:

$$S(x) = \frac{1}{1 + e^{-1}}$$

Using this function on all layers of the neural network we obtain a new learning curve, the graph of which is shown in Fig. 13. The image shows that the training was more rapid, compared with Fig. 11. For example, at the time of the 7th epoch, the network with the sigmoidal activation function gave an accuracy of approximately 37%, while

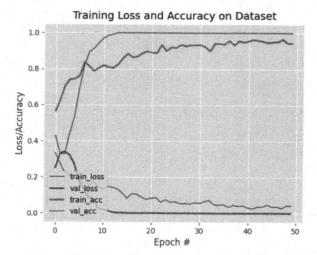

Fig. 11. Learning curve using the ReLU function

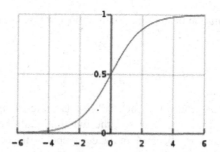

Fig. 12. Graphic representation of the sigmoid

the network with the ReLU function has already reached an accuracy of 90%. However, at the time of the 10th era, the accuracy of networks was almost equal. In further training, the network with the ReLU function behaved more stably.

This can be explained by the fact that small changes in the input data can significantly affect the outcome of sigmoid function. This has both pros and cons, because on the one hand we get more effective training in the long run, but quite unstable in the short term. As you can see, in this case, the ReLU function is more satisfactory to the conditions of the neural network, because learning is faster and more stable.

3. ELU activation function. ELU (Exponential Rectified Unit) is a function that usually brings the neural network to a faster state and gives more accurate results. Unlike other activation functions, ELU has an additional alpha constant, which must be a positive number.

ELU is very similar to ReLU, except for the behavior at negative input values. They both have the form of an identity function for non-negative inputs.

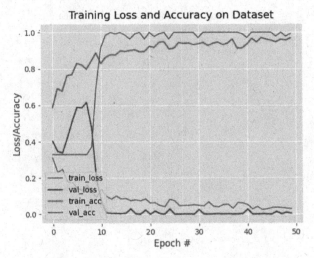

Fig. 13. Learning curve using sigmoidal function

The graphical representation of the function is shown in Fig. 14, the ELU formula is as follows:

$$ELU(x) = \begin{cases} x & if \ x > 0 \\ \alpha(e^x - 1) & if \ x \leq 0 \end{cases}$$

Using this activation function in our network we get interesting results, the learning curve is shown in Fig. 15. As we can see, the function really helped to accelerate the training of the neural network, because in the early epochs the accuracy of the network varies between 75–90%, while the network with the ReLU function at the same time of learning has an accuracy of about 35%.

4. SELU activation function. Scalable exponential linear unit - this activation func-tion is one of the newest. Its authors calculated two values: α and λ, where

$$\alpha \approx 1,6732632423543772848170429916717,$$
$$\lambda \approx 1,0507009873554804934193349852946.$$

The preset value of the parameter α means that we do not have to worry about choosing it for this activation function. In summary, SELUs have the property of self-normalization, so cases of disappearing gradients are avoided. There are three main reasons why SELU will be a good replacement for ReLU:

- Like ReLU, SELUs include deep neural networks because there are no problems with vanishing gradients.
- Unlike ReLU, SELUs cannot die, ie such a function will never reach the state when it will return the same response for any input data.
- SELUs learn on their own faster and better than other activation functions, even if they are combined with batch normalization.

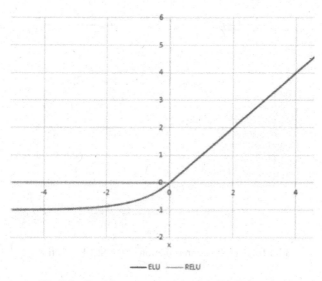

Fig. 14. Graphic representation of the ELU function

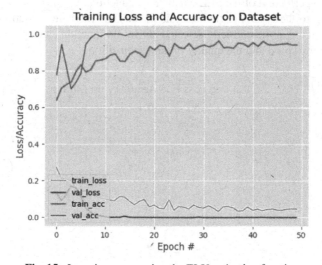

Fig. 15. Learning curve using the ELU activation function

In other respects, the function differs little from the previous one, its graphic image is shown in Fig. 16. The equation for it has the following form:

$$SELU(x) = \lambda \begin{cases} x & if \ x > 0 \\ \alpha(e^x - 1) & if \ x \leq 0 \end{cases}$$

Using the function in the developed neural network, we analyze the learning curve shown in Fig. 17. Indeed, the advantages of the function can be seen immediately, because the resulting neural network is the fastest, compared to previous experiments, came to

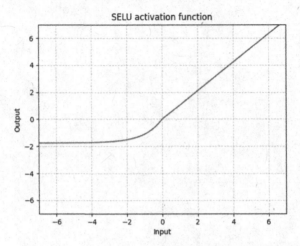

Fig. 16. Graphic representation of the SELU function

an accuracy close to 100%, with almost no fluctuations in the curve. This allows you to clearly see the perfection of the SELU activation function.

According to the results of the analysis of the four activation functions, we came to the conclusion that the correctly selected activation function can significantly affect both the learning speed and the quality of data acquisition.

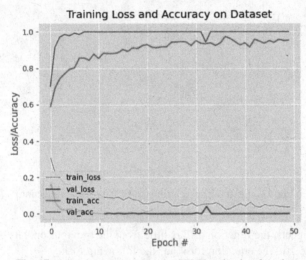

Fig. 17. Learning curve using the SELU activation function

And although the ReLU function is currently considered the best option for computer vision problems, where it is most often used, the analysis showed that the SELU function with the same training parameters showed better results, at least in the case of the developed neural network, because it led the neural network to accuracy close to 100%.

4.3 Research the Number of Layers of the Neural Network

As discussed above, each network has one input layer and one output layer. The number of neurons in the input layer is equal to the number of input variables in the data being processed. The number of neurons in the output layer is equal to the number of outputs associated with each input.

Often the problem is to determine the number of hidden layers and their neurons. The answer is that it is not possible to analytically calculate the number of layers or the number of nodes that should be used per layer in an artificial neural network to solve a specific problem of real predictive modeling.

The number of layers and the number of nodes in each layer are hyperparameters of the model that must be specified before training the network. The required amount can only be found through testing and controlled experiments.

As discussed in Sect. 3, the developed neural network model consists of a sequence of layers:

$$Convolution - Activation - Convolution - Activation - Pulling.$$

For a small neural network, the described number of layers is sufficient for the proper functioning of the network. Adding or subtracting layers can lead to network imbalances, and possible cases of oversaturation or undersaturation and the neural network model will be ineffective.

As an experiment, we add another sequence of layers of similar appearance. The dimension of the new convolution layers will now be 64, and the dimension of the last seal layer is set to 128. In Fig. 18 shows the learning curve of the network with three sequences of layers.

As reader can see, the effectiveness of learning has hardly changed, but instead of a slow increase in network accuracy, we see a sharp jump between the 10th and 15th epochs.

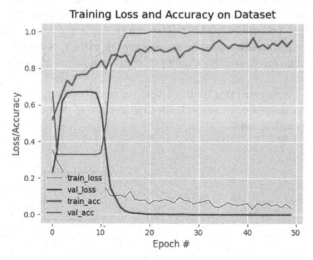

Fig. 18. Neural network learning curve with three sequences of layers

Continuing the experiments, we add another sequence of layers. Now the dimension of new layers of convolution reaches 128, and the last layer of consolidation - 256. In Fig. 19 shows the learning curve of such a model of a convolutional neural network.

Analyzing the result, you can see that the training area has decreased almost twice. So in Fig. 18 it lasted until the 10th era, and in Fig. 19 only between the 3rd and 7th epochs. Also close to 100% accuracy, the second model reached about 5 epochs faster.

So really, adding new layers to a neural network improves its learning efficiency. To consolidate the experiment, add another, last sequence of layers. The dimensions of the new convolution layers and the seal layer are now 256 and 512, respectively.

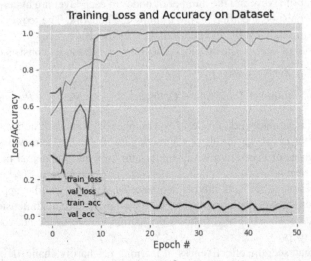

Fig. 19. Neural network learning curve with four sequences of layers

In Fig. 20 we see the result of the experiment. Although in fact the learning area has shrunk again and now runs between epochs 3 and 6, the overall stability of the curve is disturbed throughout learning. This can be explained by the phenomenon of oversaturation of the model, when the neural network extracts too many features from the input data, thus increasing the number of prediction errors and declining learning efficiency.

Thus, it was investigated that for the developed model of the neural network the use of more than four sequences of layers leads to a decrease in the quality of learning and does not make practical sense.

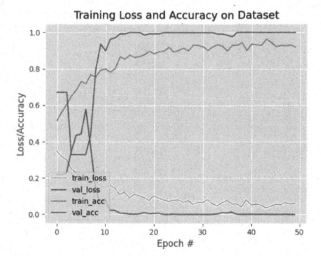

Fig. 20. Neural network learning curve with five sequences of layers

5   Conclusions

Having the results of research on the parameters, activation functions and the number of layers of the neural network, analyzing their impact on the speed and quality of learning, we can create the most efficient network.

Thus, as a result of the study of training parameters, we found that the best efficiency was provided by the value of the initial learning speed of 0.0001 and the size of the batch 10.

Among the activation functions, the most effective function was a scalable exponential linear unit, or SELU.

Considering network modifications, we also showed that four sequences of layers are sufficient for the most effective network training.

Combining the results and setting up the network, we obtain the learning curves shown in Fig. 21, 22 and 23.

As a result, we see approximately the same picture, with some error: the algorithm requires 5–8 epochs of training to work with almost 100% accuracy.

If we compare these figures with the curve shown in Fig. 6, we will see that it is almost 4 times less than it was in the basic network configuration. Moreover, the quality of assimilation has also increased significantly, because the stability of the curve is visible after 10 epochs, while the first model of the neural network such stability could not be achieved.

As a result, we see approximately the same picture, with some error: the algorithm requires 5–8 epochs of training to work with almost 100% accuracy.

If we compare these figures with the curve shown in Fig. 6, we will see that it is almost 4 times less than it was in the basic network configuration. Moreover, the quality of assimilation has also increased significantly, because the stability of the curve is visible after 10 epochs, while the first model of the neural network such stability could not be achieved.

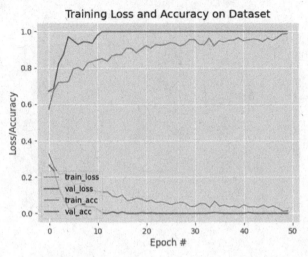

Fig. 21. Learning curves of the neural network model with the obtained parameters (launch # 1)

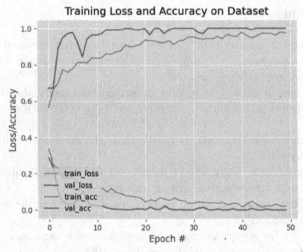

Fig. 22. Learning curves of the neural network model with the obtained parameters (launch # 2)

The studies were repeated many times, the results obtained were processed by statistical methods. The above diagrams display the average, typical result of training convolutional neural networks when processing biometric images.

Thus, we have proved that the effectiveness of learning convolutional networks depends entirely on its configuration. You can see that the graphs in Fig. 14 are somewhat unstable. To avoid this and create the most productive network, it is necessary to consider the parameters not only separately from each other, but also in the context of the network. Further modifications of both the training parameters and the layers of the

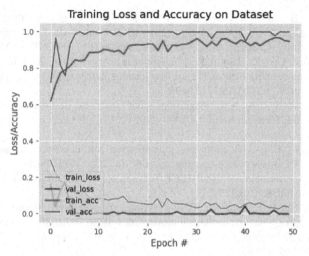

Fig. 23. Learning curves of the neural network model with the obtained parameters (launch # 3)

neural network can be performed empirically to find a combination that would provide the fastest learning with the greatest possible efficiency of learning.

References

1. Rathgeb, C., Uhl, A.: A survey on biometric cryptosystems and cancelable biometrics. EURASIP J. Inf. Secur. **2011**, 3 (2011). https://doi.org/10.1186/1687-417X-2011-3
2. Kakadiaris, I.A., Woodard, D.L., Schuckers, S.A.C.: Selected best works from biometrics: theory, applications, and systems 2019. IEEE Trans. Biom. Behav. Identity Sci. **2**, 308–309 (2020). https://doi.org/10.1109/TBIOM.2020.3022299
3. Lutsenko, M., Kuznetsov, A., Kiian, A., Smirnov, O., Kuznetsova, T.: Biometric cryptosystems: overview, state-of-the-art and perspective directions. In: Ilchenko, M., Uryvsky, L., Globa, L. (eds.) MCT 2019. LNNS, vol. 152, pp. 66–84. Springer, Cham (2021). https://doi.org/10.1007/978-3-030-58359-0_5
4. Mandal, S., Bera, B., Sutrala, A.K., Das, A.K., Choo, K.-K.R., Park, Y.: Certificateless-signcryption-based three-factor user access control scheme for IoT environment. IEEE Internet Things J. **7**, 3184–3197 (2020). https://doi.org/10.1109/JIOT.2020.2966242
5. Banerjee, S., Odelu, V., Das, A.K., Srinivas, J., Kumar, N., Chattopadhyay, S., et al.: A provably secure and lightweight anonymous user authenticated session key exchange scheme for Internet of Things deployment. IEEE Internet Things J. **6**, 8739–8752 (2019). https://doi.org/10.1109/JIOT.2019.2923373
6. Hernández Álvarez, F., Hernández Encinas, L.: Security efficiency analysis of a biometric fuzzy extractor for iris templates. In: Herrero, Á., Gastaldo, P., Zunino, R., Corchado, E. (eds.) Computational Intelligence in Security for Information Systems, pp. 163–170. Springer, Heidelberg (2009). https://doi.org/10.1007/978-3-642-04091-7_20
7. Shreyas, K.K.M., Rajeev, S., Panetta, K., Agaian, S.S.: Fingerprint authentication using geometric features. In: 2017 IEEE International Symposium on Technologies for Homeland Security (HST), pp. 1–7 (2017). https://doi.org/10.1109/THS.2017.7943449

8. Perazzone, J.B., Yu, P.L., Sadler, B.M., Blum, R.S.: Fingerprint embedding authentication with artificial noise: MISO regime. In: 2019 IEEE Conference on Communications and Network Security (CNS), pp. 1–5 (2019). https://doi.org/10.1109/CNS.2019.8802636

9. Engelsma, J.J., Cao, K., Jain, A.K.: Fingerprint match in box. In: 2018 IEEE 9th International Conference on Biometrics Theory, Applications and Systems (BTAS), pp. 1–10 (2018). https://doi.org/10.1109/BTAS.2018.8698556

10. Chugh, T., Cao, K., Jain, A.K.: Fingerprint spoof buster: use of minutiae-centered patches. IEEE Trans. Inf. Forensics Secur. **13**, 2190–2202 (2018). https://doi.org/10.1109/TIFS.2018.2812193

11. Chuang, C.-W., Fan, C.-P.: Biometric authentication with combined iris and sclera information by YOLO-based deep-learning network. In: 2020 IEEE International Conference on Consumer Electronics - Taiwan (ICCE-Taiwan), pp. 1–2 (2020). https://doi.org/10.1109/ICCE-Taiwan49838.2020.9258253

12. Hsiao, C.-S., Fan, C.-P., Hwang, Y.-T.: Iris location and recognition by deep-learning networks based design for biometric authorization. In: 2021 IEEE 3rd Global Conference on Life Sciences and Technologies (LifeTech), pp. 144–145 (2021). https://doi.org/10.1109/LifeTech52111.2021.9391787

13. Boutros, F., Damer, N., Raja, K., Ramachandra, R., Kirchbuchner, F., Kuijper, A.: On benchmarking iris recognition within a head-mounted display for AR/VR applications. In: 2020 IEEE International Joint Conference on Biometrics (IJCB), pp. 1–10 (2020). https://doi.org/10.1109/IJCB48548.2020.9304919

14. Oldal, L.G., Kovács, A.: Hand geometry and palmprint-based authentication using image processing. In: 2020 IEEE 18th International Symposium on Intelligent Systems and Informatics (SISY), pp. 125–130 (2020). https://doi.org/10.1109/SISY50555.2020.9217068

15. Gupta, P., Gupta, P.: Multibiometric authentication system using slap fingerprints, palm dorsal vein, and hand geometry. IEEE Trans. Ind. Electron. **65**, 9777–9784 (2018). https://doi.org/10.1109/TIE.2018.2823686

16. Kinkiri, S., Keates, S.: Speaker identification: variations of a human voice. In: 2020 International Conference on Advances in Computing and Communication Engineering (ICACCE), pp. 1–4 (2020). https://doi.org/10.1109/ICACCE49060.2020.9154998

17. Miguel-Hurtado, O., Blanco-Gonzalo, R., Guest, R., Lunerti, C.: Interaction evaluation of a mobile voice authentication system. In: 2016 IEEE International Carnahan Conference on Security Technology (ICCST), pp. 1–8 (2016). https://doi.org/10.1109/CCST.2016.7815697

18. Miguel-Hurtado, O., Guest, R., Lunerti, C.: Voice and face interaction evaluation of a mobile authentication platform. In: 2017 International Carnahan Conference on Security Technology (ICCST), pp. 1–6 (2017). https://doi.org/10.1109/CCST.2017.8167860

19. Face-based multiple user active authentication on mobile devices (n.d.). https://ieeexplore.ieee.org/document/8501956/. Accessed 8 May 2021

20. A face-recognition approach using deep reinforcement learning approach for user authentication (n.d.). https://ieeexplore.ieee.org/document/8119148/. Accessed 8 May 2021

21. Beukes, E., Coetzer, J.: Hand vein-based biometric authentication using two-channel similarity measure networks. In: 2020 International SAUPEC/RobMech/PRASA Conference, pp. 1–6 (2020). https://doi.org/10.1109/SAUPEC/RobMech/PRASA48453.2020.9041064

22. Kohlakala, A., Coetzer, J.: On automated ear-based authentication. In: 2020 International SAUPEC/RobMech/PRASA Conference, pp. 1–6 (2020). https://doi.org/10.1109/SAUPEC/RobMech/PRASA48453.2020.9041089

23. Karimian, N., Woodard, D., Forte, D.: ECG biometric: spoofing and countermeasures. IEEE Trans. Biom. Behav. Identity Sci. **2**, 257–270 (2020). https://doi.org/10.1109/TBIOM.2020.2992274

24. Mondal, S., Bours, P.: Swipe gesture based continuous authentication for mobile devices. In: 2015 International Conference on Biometrics (ICB), pp. 458–465 (2015). https://doi.org/10.1109/ICB.2015.7139110

25. Mowla, N., Doh, I., Chae, K.: Selective fuzzy ensemble learner for cognitive detection of bio-identifiable modality spoofing in MCPS. In: 2018 20th International Conference on Advanced Communication Technology (ICACT), pp. 63–67 (2018). https://doi.org/10.23919/ICACT.2018.8323646

26. Vareto, R.H., Saldanha, A.M., Schwartz, W.R.: The swax benchmark: attacking biometric systems with wax figures. In: ICASSP 2020 - 2020 IEEE International Conference on Acoustics, Speech and Signal Processing (ICASSP), pp. 986–990 (2020). https://doi.org/10.1109/ICASSP40776.2020.9053946

27. Lee, C.E., Zheng, L., Zhang, Y., Thing, V.L.L., Chu, Y.Y.: Towards building a remote anti-spoofing face authentication system. In: TENCON 2018 - 2018 IEEE Region 10 Conference, pp. 0321–0326 (2018). https://doi.org/10.1109/TENCON.2018.8650440

28. Matthew, P., Anderson, M.: Developing coercion detection solutions for biometrie security. In: 2016 SAI Computing Conference (SAI), pp. 1123–1130 (2016). https://doi.org/10.1109/SAI.2016.7556118

29. Wu, L., Yang, J., Zhou, M., Chen, Y., Wang, Q.: LVID: a multimodal biometrics authentication system on smartphones. IEEE Trans. Inf. Forensics Secur. **15**, 1572–1585 (2020). https://doi.org/10.1109/TIFS.2019.2944058

30. Aware Logo: Liveness detection in biometrics is essential for mobile authentication and onboarding. Aware (2019). https://www.aware.com/blog-liveness-detection-mobile-authentication-onboarding/. Accessed 12 Feb 2021

31. Qayyum, A.B.A., Arefeen, A., Shahnaz, C.: Convolutional neural network (CNN) based speech-emotion recognition. In: 2019 IEEE International Conference on Signal Processing, Information, Communication Systems (SPICSCON), pp. 122–125 (2019). https://doi.org/10.1109/SPICSCON48833.2019.9065172

32. Lou, G., Shi, H.: Face image recognition based on convolutional neural network. China Commun. **17**, 117–124 (2020). https://doi.org/10.23919/JCC.2020.02.010

33. McCulloch, W.S., Pitts, W.: A logical calculus of the ideas immanent in nervous activity. Bull. Math. Biophys. **5**, 115–133 (1943). https://doi.org/10.1007/BF02478259

34. Farley, B., Clark, W.: Simulation of self-organizing systems by digital computer. Trans. IRE Prof. Group Inf. Theory **4**, 76–84 (1954). https://doi.org/10.1109/TIT.1954.1057468

35. Ahmed, T., Das, P., Ali, M.F., Mahmud, M.-F.: A comparative study on convolutional neural network based face recognition. In: 2020 11th International Conference on Computing, Communication and Networking Technologies (ICCCNT), pp. 1–5 (2020). https://doi.org/10.1109/ICCCNT49239.2020.9225688

36. Hebb, D.O.: The Organization of Behavior: A Neuropsychological Theory. Psychology Press, New York (2005)

37. Qin, Z., Huang, G., Xiong, H., Qin, Z., Choo, K.-K.R.: A fuzzy authentication system based on neural network learning and extreme value statistics. IEEE Trans. Fuzzy Syst. **29**, 549–559 (2021). https://doi.org/10.1109/TFUZZ.2019.2956896

38. Kuznetsov, A., Oleshko, I., Chernov, K., Bagmut, M., Smirnova, T.: Biometric authentication using convolutional neural networks. In: Ilchenko, M., Uryvsky, L., Globa, L. (eds.) MCT 2019. LNNS, vol. 152, pp. 85–98. Springer, Cham (2021). https://doi.org/10.1007/978-3-030-58359-0_6

39. Pinto, A., Schwartz, W.R., Pedrini, H., de Rezende Rocha, A.: Using visual rhythms for detecting video-based facial spoof attacks. IEEE Trans. Inf. Forensics Secur. **10**, 1025–1038 (2015). https://doi.org/10.1109/TIFS.2015.2395139

40. Rattani, A., Scheirer, W.J., Ross, A.: Open set fingerprint spoof detection across novel fabrication materials. IEEE Trans. Inf. Forensics Secur. **10**, 2447–2460 (2015). https://doi.org/10.1109/TIFS.2015.2464772

41. Patel, K., Han, H., Jain, A.K.: Secure face unlock: spoof detection on smartphones. IEEE Trans. Inf. Forensics Secur. **11**, 2268–2283 (2016). https://doi.org/10.1109/TIFS.2016.2578288

42. Wen, D., Han, H., Jain, A.K.: Face spoof detection with image distortion analysis. IEEE Trans. Inf. Forensics Secur. **10**, 746–761 (2015). https://doi.org/10.1109/TIFS.2015.2400395

43. Nikisins, O., Mohammadi, A., Anjos, A., Marcel, S.: On effectiveness of anomaly detection approaches against unseen presentation attacks in face anti-spoofing. In: 2018 International Conference on Biometrics (ICB), pp. 75–81 (2018). https://doi.org/10.1109/ICB2018.2018.00022

44. Pinto, A., Pedrini, H., Schwartz, W.R., Rocha, A.: Face spoofing detection through visual codebooks of spectral temporal cubes. IEEE Trans. Image Process. **24**, 4726–4740 (2015). https://doi.org/10.1109/TIP.2015.2466088

45. Rathgeb, C., Wagner, J., Tams, B., Busch, C.: Preventing the cross-matching attack in Bloom filter-based cancelable biometrics. In: 3rd International Workshop on Biometrics and Forensics (IWBF 2015), pp. 1–6 (2015). https://doi.org/10.1109/IWBF.2015.7110226

46. Steiner, H., Kolb, A., Jung, N.: Reliable face anti-spoofing using multispectral SWIR imaging. In: 2016 International Conference on Biometrics (ICB), pp. 1–8 (2016). https://doi.org/10.1109/ICB.2016.7550052

47. Chugh, T., Jain, A.K.: Fingerprint spoof detector generalization. IEEE Trans. Inf. Forensics Secur. **16**, 42–55 (2021). https://doi.org/10.1109/TIFS.2020.2990789

48. Edwards, T., Hossain, M.S.: Effectiveness of deep learning on serial fusion based biometric systems. IEEE Trans. Artif. Intell. 1 (2021). https://doi.org/10.1109/TAI.2021.3064003

49. IEEE Standard for Biometric Liveness Detection. IEEE Std 2790-2020, pp. 1–24 (2020). https://doi.org/10.1109/IEEESTD.2020.9080669

50. Babu, A., Paul, V., Baby, D.E.: An investigation of biometric liveness detection using various techniques. In: 2017 International Conference on Inventive Systems and Control (ICISC), pp. 1–5 (2017). https://doi.org/10.1109/ICISC.2017.8068745

51. Okereafor, K., Onime, C., Osuagwu, O.: Enhancing biometric liveness detection using trait randomization technique. In: 2017 UKSim-AMSS 19th International Conference on Computer Modelling Simulation (UKSim), pp. 28–33 (2017). https://doi.org/10.1109/UKSim.2017.44

52. Garg, S., Mittal, S., Kumar, P., Athavale, V.A.: DeBNet: multilayer deep network for liveness detection in face recognition system. In: 2020 7th International Conference on Signal Processing and Integrated Networks (SPIN), pp. 1136–1141 (2020). https://doi.org/10.1109/SPIN48934.2020.9070853

53. Liveness Detection with OpenCV. PyImageSearch (2019). https://www.pyimagesearch.com/2019/03/11/liveness-detection-with-opencv/. Accessed 12 Feb 2021

54. Chingovska, I., Anjos, A., Marcel, S.: On the effectiveness of local binary patterns in face anti-spoofing. In: 2012 BIOSIG - Proceedings of the International Conference of Biometrics Special Interest Group (BIOSIG), pp. 1–7 (2012)

Process Mining for the Analysis of Pre-sales Customer Service Process – A Hidden Observation in a Polish Automotive Organization

Piotr Sliż[(✉)] [iD] and Emilia Dobrowolska[iD]

Uniwersytet Gdański, Wydział Zarządzania, ul. Armii Krajowej 101, 81-824 Sopot, Poland
{piotr.sliz,emilia.dobrowolska}@ug.edu.pl

Abstract. The main goal of the article was to present the pre-sales of the customer service process using process mining, i.e. assessing the course of activities related to establishing contact with the customer and presenting the sale offer. As a result of the completed proceedings, it was noticed that the examined pre-sales process requires optimization due to such parameters as: long time of implementation of activities in the process, failure to take actions aimed at presenting the offer and incompatible with the clients inquiry presentation. The proceedings described in this article proved that it is possible to analyse the pre-sales process described using the combination of hidden non-participant observation methods and process mining. Unlike expost studies based on data provided by the surveyed organizations, the proposed solution eliminates errors related to data quality, but it is prone to errors related to IT infrastructure, which include problems with e-mail recipients or problems with delivery of messages.

Keywords: Process-mining · Process orientation · BPO · BPM · Sales process · Automotive

JEL: M210

1 Introduction

The dynamic development of technologies related to electromobility in the business ecosystems of modern organizations generates a number of changes implying the need to reconfigure business models, production processes, but also organizational structures of enterprises from the automotive sector. In the space of implementation of modern technical solutions, enabling the increase in the number of sales and the share of cars with alternative energy sources for combustion engines, by far the most developed countries in Europe are Norway, Iceland, Sweden and the Netherlands [32]. This is also noticeable in Poland, which, from a political, legal and social perspective is slowly becoming a stakeholder of the global change.

Solutions, related to the electrification of passenger cars, bring a number of changes for organizations participating in the discussed sector, starting with manufacturers,

S. Wrycza and J. Maślankowski (Eds.): PLAIS EuroSymposium 2021, LNBIP 429, pp. 147–160, 2021.
https://doi.org/10.1007/978-3-030-85893-3_10

importers, part manufactures, networks of authorized dealerships, ending with enterprises supplying electricity and designing and providing infrastructure related to servicing and charging this kind of vehicles. To sum up, implementing new techniques and technologies becomes a catalyst for the development of the digital economy, which is an indispensable element in the organization's space, these are ICT innovation which have a positive impact on increasing the effectiveness of the organization and the activities carried out in it [9].

This article assumes that business processes constituting the core of the organization [30] occur in each of them [5], and their effects (products and services) are customer-oriented in external and internal terms. Therefore, modern managers should take the challenge aimed at increasing the level of process orientation, to generate a system state in which the organization will continuously discount the benefits of achieving subsequent levels of process maturity. In age of digitizing economy, this requires exploration of data generated inside organizations, but also in its surroundings. One solution to this problem is the use of process mining, which in the literature is defined as "a relative young research discipline that sits between machine learning and data mining on the one hand and process modelling and analysis on the other hand. (…) The idea of process mining is to discover, monitor and improve real processes (i.e., not assumed processes) by extracting knowledge from event logs readily available in today's systems" [23]. According to the authors, the use of the process mining method should be considered useful in assessing the main processes for which external customer service has been qualified, to verify the actual course of the process and to optimize it, based on the results of data mining, by reducing activities that do not provide added value.

The main goal of the article was to present the pre-sales of the customer service process using process mining, i.e. assessing the course of activities related to establishing contact with the customer and presenting the sale offer. Implementation of the main goal was assigned with auxiliary goals, interpenetrating in the epistemological (EG) and methodical (MG) dimensions. EG: Approximation and determination of the current state of knowledge regarding process mining and its application in research with a high degree of application in business practice. MG: Presentation of a proposal for the methodology of reconstruction of pre-sales business process using the process mining method.

The following research methods were used to achieve the stated goals: quantitative bibliometric analysis, systematic literature review, non-participant implicit (hidden) observation, process-mining and statistical methods. The analysis of the results obtained was carried out using the Celonis program and the R programming language.

The first section of the article presents the level of interest and areas of process-mining implementation using bibliometric analysis and a systematic review of the literature. The next section summarizes the interest in process-mining in the field of science and business. Subsequent section presents the course of the research process, including the description of the research procedure, characteristics of the tested units and data. The Celonis SNAP program enabled the authors to present the results of the study in the following section of the article. The last section summarizes the obtained results, outlines the limitations of the study and outlines the directions for further research.

2   Application of Process Mining – Assessment of the State of Knowledge and Identification of Cognitive Gaps

2.1   Quantitative Bibliometric Analysis

The purpose of the theoretical study was to formulate cognitive gaps based on which the goals and research question were formulated. Table 1 presents the summary of the number of publications and the number of citations for the examined entry ('process mining') based on the Web of Science (WoS) database.

Table 1. List of bibliometric indicators for the entry "process mining" in the WOS database

Entry/entries	Database	Years of publications	Total publications	Sum of times cited	Citing articles	Average citations per item	h-index
"process mining" OR "process mining"	Web of Science	2003–2020	654	7313*/5202**	3491*/2953**	11,18	38

*Including auto-citations
**No auto-citations.
Source: own study based on the WoS database, access: 08.01.2020.

In turn, Fig. 1 shows detailed data on both studied parameters.

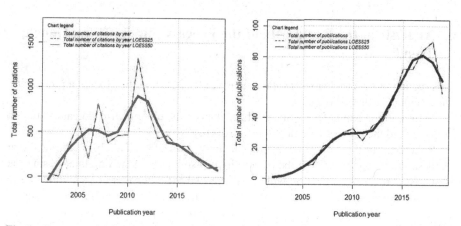

Fig. 1. Summary of the number of publications and the number of their citations in 2003–2020 for the entry 'process mining' based on the WoS database. Source: own study using the R programming language based on the WoS database, access: 08.01.2020.

The analysis of the time series presented in Fig. 1 was performed using non-parametric locally weighted regression identified in the literature with the abbreviation LOESS [2]. The possibilities, advantages and limitations of using LOESS regression are described in detail in [14].

2.2 Systematic Review of the Literature on the Subject

This section of the article focuses primarily on the description and assessment of the spectrum of use of process mining in organization management and the possibilities of its use in assessing the course of the customer service process. The systematic literature review made it possible to outline the areas in which the process mining is used and is of an application nature. At this point, it should be emphasized that the purpose of this article was not to characterize the concept of the process mining, which was precisely explained in the literature on the subject [25, 28, 29].

The literature review provided evidence that the process mining has been widely used in the analysis of the health care sector processes, with particular emphasis on health care environments [8, 17, 18], as well as in the narrower scope on the example of selected health care units [12, 16] and the analysis of selected medical processes [4, 21]. The studies attempted to detect the health care fraud and abuse [31], and also performed detailed medical analyses, such as: an application to stroke care [13]. In broader terms, without indicating the sector, the process mining was of interest to researchers in such areas as: anomaly detection [1], auditing [6], time prediction [26], internal transaction fraud mitigation [7], supporting the work of auditors [27] and service behavior discover, check and improve (service mining) [24]. From the perspective of the automotive sector analyzed units, the interest of researchers in the application of the process mining in the after-sales service process [22] and in car production processes [3] has been identified. In turn, the use of the process mining on the level of assessing customer relations has been presented in the following issues: analysis of customer fulfilment [11, 15], analysis of customers journey [19, 20] and customer integration in service processes [10].

3 Analysis of the Popularity of the Process Mining Using Google Trends

This section of the article presents the interest in the issues of the process mining using the Google Trends tool (Fig. 2).

In turn, Fig. 3 compares the standardized data on the bibliometric indicators studied (Table 1) and the level of Google interest in 2002−2019.

As a result of the completed theoretical study, it should be clearly emphasized that the described issues regarding the possibility of using process mining in assessing the relationship of the organization with the client and the assessment of activities in the course of processes are current in the space of researchers presenting the results of considerations and research in the literature and business practice. This is indicated by the following premises:

- process mining is interdisciplinary, as evidenced by document analysis according to the web of science categories. The entry of process mining was most often mentioned in publications in WoS research areas, such as: computer science (78.135%), engineering (24.465%), operations research management science (7.951%) as well: telecommunications, business economics and medical informatics (<5%);

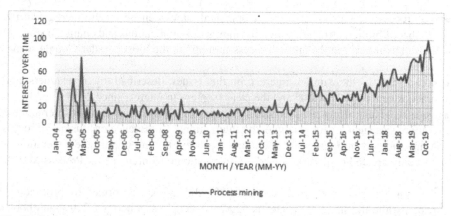

Fig. 2. Summary of the interest of the 'process mining' entry using the Google Trends tool. *Category: all, assessment range: worldwide. Source: own study based on Google Trends (www. trends.google.com), access: 08.01.2020.

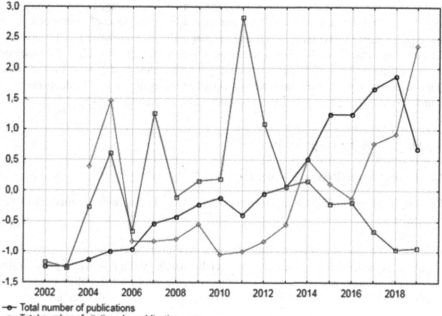

-●- Total number of publications
-●- Total number of citations by publication year
-◆- Google search interest

Fig. 3. Summary of standardized data on the total number of publications, citations by year, interest in Google search engine in 2002–2019*. *For the Google search interest variable, due to the lack of data, the analyzed period includes years 2004–2019. Source: own study based on Google Trends (www.trends.google.com), access: 08.01.2020.

- the process mining issues are developmental, as evidenced by the increase in the number of bibliometric indicators (the number of publications and citations) (Fig. 1), as well as interest in the entry "process-mining" in the most popular Google search engine (Fig. 2);
- noteworthy is the similar interest in the issues described by researchers and practitioners in 2004–2005 and the clear increase in interest since 2013 (Fig. 3);
- research results using process mining are presented by researchers at international conferences such as, for example, the International Conference on Business Process Management, and in addition, the issue of process mining implies the generation of specially created conferences, like the International Conference on Process Mining ICPM.
- the vast majority, as much as 65.902% of publications are formed by proceedings paper, while the share of articles is 32.875%, chapters in books 2.299% and books 0.459%[1].

As a result of the theoretical study carried out in the first section of the article, a cognitive gap was identified, it consists of a small number of publications regarding the possibility of using the process mining method in the assessment of the client-organization relationship and the lack of research on the analysis of the course of customer service based on the event log registered by the researcher using the non-participant observation method. Due to the limited access to data generated in the organization, low response rate in surveys and problems related to data quality noticed in the discipline of management and quality sciences, it can generate new possibilities in the area of evaluation of the course of main processes oriented at external clients in organizations.

4   Materials and Methods

4.1   Project of the Research Procedure

The empirical proceedings were carried out in July 2019 and were divided into three successive stages.

- Stage 1 – Preparation of the register of examined units based on information available on the website of the Automotive Market Research Institute – Samar.pl, websites of manufacturers and dealerships offering electric cars on the Polish market. On this basis, a register of e-mail contact addresses for new car sales departments was prepared.
- Stage 2 – An e-mail was sent to each of the surveyed organizations at the same time, they were asked to submit an offer about the available models, versions and catalogue prices of electric cars. In addition, no selected source of financing was indicated with a request to present its various options. The last question considered the matter whether the offer includes ready-to-pick cars, and in the case of a negative answer, what is the waiting period from the moment of placing the order.

---

[1] Data formulated on the basis of the quantitative bibliometric analysis based on the Web of Science database, access 08.01.2020.

- Stage 3 – At this stage, a conditioning criterion was prepared, i.e. after fulfilling the criterion of answering all questions from the first e-mail, supplementary questions were sent regarding: availability of the warranty service at the dealership, warranty period, costs and time period for performing the review. In addition, dealerships were asked about the availability of maps with chargers and the possibility of charging the vehicle at home.

The detailed course of the research process is shown in Fig. 4. The purpose of the designed procedure map was to replicate the procedure.

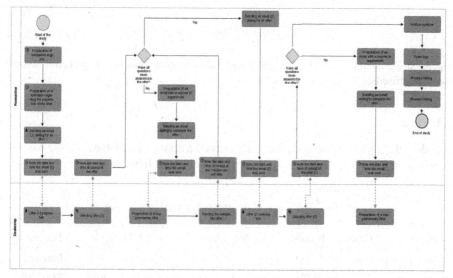

Fig. 4. Map of the course of activities in the research process implemented in 2019. Source: own study using the BOC Adonis program.

4.2   Characteristics of the Research Sample

The research was carried out on a complete sample of 341 authorized service stations for passenger cars in Poland, belonging to the dealership network of car manufacturers, which during the study offered electric cars on the Polish market. The structure of the surveyed units in terms of number and brand is shown in Fig. 5.

At this point, it should be emphasized that in 2020 the number of offered brands and models with an electric drive on the Polish market increased, which is the reason to continue the next study. In addition, it was noticed during the empirical proceedings that the sale of electric cars could not be carried out by all authorized dealerships, although they belong to the dealership network of the manufacturer that offers such cars. It should be noted that such information was not presented on all official websites of manufacturers and authorized stations in Poland.

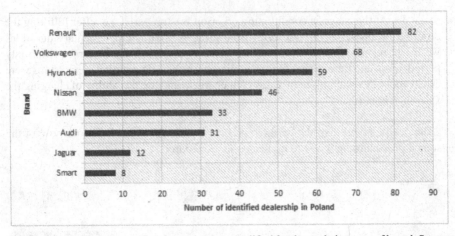

Fig. 5. Summary of the number of organizations qualified for the study in terms of brand. Source: own study based on the study performed in 2019.

4.3 Data

Based on the obtained raw data, the study created a relational database, on the basis of which event logs were designed. An example is shown in Table 2.

Table 2. Structure of the events-log used in the study

Variable	CaseID	Brand	Dealership	User	Activity	Timestamp
Variable class	*float*	*string*	*string*	*string*	*string*	*Datetime*
Activity_a	329	BMW	D_Pol_329	Customer	Offer request – first contact	24-07-19 7:30
Activity_b	329	BMW	D_Pol_329	Sales advisor	Presentation of the offer - first contact	24-07-19 10:25
Activity_c	329	BMW	D_Pol_329	Customer	Offer request – second contact	25-07-19 12:51
Activity_d	329	BMW	D_Pol_329	Sales advisor	Reply to second contact	25-07-19 14:18
Activity_e	329	BMW	D_Pol_329	Sales advisor	Reply to third contact	30-07-19 14:42
Activity_f	329	BMW	D_Pol_329	Customer	Offer request – third contact	30-07-19 14:55

Source: own study based on a study completed in 2018.

The event logs database was prepared in accordance with the recommendations of the Celonis program producer, with the help of which the examined process was explored. At this point, it should be emphasized that the lack of the reference model of the studied pre-sales process in authorized service stations offering electric cars was the limitation in the study. Presales process is defined as a set of consecutive activities focused on an external customer whose output is the input to the main sales and customer service process. The purpose of this pre-sales process is the transformation of queries and individualized requirements of the customer regarding the product or service trade offer.

5 Results and Discussion

As a result of the process analysis using the Celonis program, 329 cases and 689 events were identified. It should be understood that a message with a question request was sent to the manufacturer who offers the sale of electric cars for all identified dealerships that meet the criterion of having a license. Figure 6 presents the happy path algorithm for the examined process as the flow from most frequent starting activity to the most frequent ending activity that was identified only for 5.01% cases.

Fig. 6. Algorithmic happy path. Source: own study based on the study carried out in 2019 using the Celonis program.

Figure 7 presents the process model taking into account all 339 cases examined, while Fig. 8 the model for the biggest cluster for 74.60% of the analysed cases.

As a result, the following analysis parameters were presented through the Celonis program. First of all, the total number of cases per day = 33 cases and the total number of events per day = 68 events were identified during the study. In addition, it was estimated that the average case duration from process start to end without extreme outliers = 56 h, while the average time was 64 h. With reference to this value, the parameter of happy path throughout time amounted to 93 versus 56 h. The analysis of the model presented in Fig. 7 makes it possible to assess how a small number of organizations responded to the offer request.

At this point, it should be emphasized that in order to improve the quality of the obtained data, each e-mail was verified in details.

The model in Fig. 8 identifies activities that occur at a high frequency, but are not necessarily part of the optimal process.

The activities that do not bring added value for an external customer include: automatic reply (in 21% of cases, 70 events), sending partial offer (in 14% of cases, 46%) and sending a request to complete offer (in 13% of cases, 43 events).

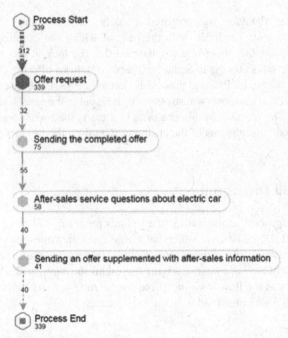

Fig. 7. Process model for all 339 cases. Source: own study based on the study carried out in 2019 using the Celonis program.

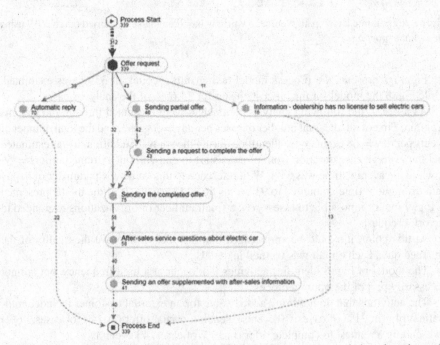

Fig. 8. Process model from the biggest cluster (74.60%). Source: own study based on the study carried out in 2019 using the Celonis program.

6   Conclusion

Summing up the obtained results of the analysis of the examined pre-sales customer service process, four general conclusions can be formulated.

First of all, the proceedings described in this article proved that it is possible to analyze the pre-sales process described using the combination of hidden non-participant observation methods and process mining. Unlike ex-post studies based on data provided by the surveyed organizations, the proposed solution eliminates errors related to data quality, but is prone to errors related to IT infrastructure, which include problems with e-mail recipients or problems with delivery of messages.

Secondly, the study has shown that, in the case of a larger number of organizations, sellers gave up on answering the question or the response time using e-mail was too long (the average time of the examined pre-sales process is 64 h).

Thirdly, it is worth noting that the sales department employees who carried out activities in the described process were well acquainted with the after-sales service issues. This was verified based on a small number of message redirections to service department employees. The vast majority of responses received were forwarded by the sales department employee.

The fourth and final conclusion is that the study did not identify any significant differences in the implementation of the process under examination due to the dealership brand.

7   Limitations of the Completed Study

In the first quarter of 2020 the share of manufacturers and the number of offered electric car models increased significantly on the Polish market. On the one hand, this is a limitation resulting from the inability to present the current state of the customer process service, but on the other hand it is an incentive to undertake further research based on the reference model of the process, built on the basis of the conclusions formulated in this study.

8   Direction of Further Research

The authors goal is to repeat the study using the method of research experiment, which, unlike observations, will be characterized by changing parameters during the implementation of empirical proceedings. As a result of the proceedings and the outline of restrictions, the direction of future research was formulated, which was to re-implement the research on an expanded dealer base in Poland with an extended mileage assessment scale. In addition, the goal is to carry out a similar international study, in particular in countries where the volume of sales of electric cars is higher than in Poland, such as Norway, Sweden, Germany and the United Kingdom were qualified to them.

9 Practical Implication

The presented solution enabling the reconstruction of the described pre-sales customer service process on the ICT path indicates the possibility of wide application of the presented approach using the process mining method. First of all, it enables dynamic assessment of authorized service stations in the area of customer service, giving up the methods of opinion polls. Secondly, from the perspective of a process approach in management, it enables the assessment of customer orientation including electronic contact as the main contract channel. Thirdly, a broad analysis of events in the described process may have a positive impact on the effects of designing standards related to the implementation of customer service activities for licensors.

10 Summary

The development of modern technologies, as well as information and communication technique, determines the level of digitalization of the modern economy. The automotive sector is one of its branches in which this can be seen. This is especially noticeable in the area of progress in the processes of production, sale and rental of electric cars. The authors formulated a research question about the pre-sales process, from submitting a request question via e-mail to its final presentation in the customer service process. The results of the study show the possibilities of conducting business process analysis using the method of non-participatory observation and process mining. As a result of the completed proceedings, it was noticed that the examined pre-sales process requires optimization due to such parameters as: long time of implementation of activities in the process, failure to take actions aimed at presenting the offer and presentation of the offer incompatible with the client's inquiry. In addition, the described results and conclusions form a recommendation for organizations operating in the automotive sector, both for authorized stations that can optimize the customer service process, but also for car manufacturers and importers who implement the concepts of customer satisfaction assessment, replacing the mysterious customer used in the examined sector using the method with the application of process mining.

References

1. Bezerra, F., Wainer, J., van der Aalst, W.M.P.: Anomaly detection using process mining. In: Halpin, T., et al. (eds.) BPMDS/EMMSAD -2009. LNBIP, vol. 29, pp. 149–161. Springer, Heidelberg (2009). https://doi.org/10.1007/978-3-642-01862-6_13
2. Cleveland, W.S., Devlin, S.J.: Locally weighted regression: an approach to regression analysis by local fitting. J. Am. Stat. Assoc. **83**, 596–610 (1988). https://doi.org/10.1080/01621459.1988.10478639
3. Dišek, M., Šperka, R., Kolesár, J.: Conversion of real data from production process of automotive company for process mining analysis. In: Jezic, G., Kusek, M., Chen-Burger, Y.-H., Howlett, R.J., Jain, L.C. (eds.) KES-AMSTA 2017. SIST, vol. 74, pp. 223–233. Springer, Cham (2018). https://doi.org/10.1007/978-3-319-59394-4_22

4. Fernandez-Llatas, C., Lizondo, A., Monton, E., Benedi, J.-M., Traver, V.: Process mining methodology for health process tracking using real-time indoor location systems. Sensors. **15**, 29821–29840 (2015). https://doi.org/10.3390/s151229769
5. Hammer, M., Stanton, S.: How process enterprises really work. Harv. Bus. Rev. **77**, 108–120 (1999)
6. Jans, M., Alles, M.G., Vasarhelyi, M.A.: A field study on the use of process mining of event logs as an analytical procedure in auditing. Account. Rev. **89**, 1751–1773 (2014). https://doi.org/10.2308/accr-50807
7. Jans, M., van der Werf, J.M., Lybaert, N., Vanhoof, K.: A business process mining application for internal transaction fraud mitigation. Expert Syst. Appl. **38**, 13351–13359 (2011). https://doi.org/10.1016/j.eswa.2011.04.159
8. Kaymak, U., Mans, R., van de Steeg, T., Dierks, M.: On process mining in health care. In: 2012 IEEE International Conference on Systems, Man, and Cybernetics (SMC), Seoul, Korea (South), pp. 1859–1864. IEEE (2012)
9. Kiełtyka, L., Kobis, P.: Ekonomiczne aspekty wirtualizacji zasobów informatycznych przedsiębiorstw. PO 13–19 (2013). https://doi.org/10.33141/po.2013.04.03
10. Leyer, M., Moormann, J.: Combining process mining and statistical methods to evaluate customer integration in service processes. In: Daniel, F., Barkaoui, K., Dustdar, S. (eds.) BPM 2011. LNBIP, vol. 99, pp. 147–152. Springer, Heidelberg (2012). https://doi.org/10.1007/978-3-642-28108-2_14
11. Mahendrawathi, E.R., Astuti, H.M., Nastiti, A.: Analysis of customer fulfilment with process mining: a case study in a telecommunication company. Procedia Comput. Sci. **72**, 588–596 (2015). https://doi.org/10.1016/j.procs.2015.12.167
12. Mans, R.S., Schonenberg, M.H., Song, M., van der Aalst, W.M.P., Bakker, P.J.M.: Application of process mining in healthcare – a case study in a Dutch hospital. In: Fred, A., Filipe, J., Gamboa, H. (eds.) BIOSTEC 2008. CCIS, vol. 25, pp. 425–438. Springer, Heidelberg (2008). https://doi.org/10.1007/978-3-540-92219-3_32
13. Mans, R., et al.: Process mining techniques: an application to stroke care. In: MIE, vol. 136, pp. 425–438 (2008)
14. Poniat, R.: On the possibility of using the LOESS regression in the analysis of time series. Przeszłość Demograficzna Polski **38**, 104–115 (2016). https://doi.org/10.18276/pdp.2016.2.38-04
15. R'bigui, H., Cho, C.: Customer oder fulfillment process analysis with process mining: an industrial application in a heavy manufacturing company. In: Proceedings of the 2017 International Conference on Computer Science and Artificial Intelligence - CSAI 2017, Jakarta, Indonesia, pp. 247–252. ACM Press (2017)
16. Rebuge, Á., Ferreira, D.R.: Business process analysis in healthcare environments: a methodology based on process mining. Inf. Syst. **37**, 99–116 (2012). https://doi.org/10.1016/j.is.2011.01.003
17. Rojas, E., Munoz-Gama, J., Sepúlveda, M., Capurro, D.: Process mining in healthcare: a literature review. J. Biomed. Inform. **61**, 224–236 (2016). https://doi.org/10.1016/j.jbi.2016.04.007
18. Rovani, M., Maggi, F.M., de Leoni, M., van der Aalst, W.M.P.: Declarative process mining in healthcare. Expert Syst. Appl. **42**, 9236–9251 (2015). https://doi.org/10.1016/j.eswa.2015.07.040
19. Terragni, A., Hassani, M.: Analyzing customer journey with process mining: from discovery to recommendations. In: 2018 IEEE 6th International Conference on Future Internet of Things and Cloud (FiCloud), Barcelona, Spain, pp. 224–229. IEEE (2018)
20. Terragni, A., Hassani, M.: Optimizing customer journey using process mining and sequence-aware recommendation. In: Proceedings of the 34th ACM/SIGAPP Symposium on Applied Computing, Limassol Cyprus, pp. 57–65. ACM (2019)

21. Tsumoto, S., Iwata, H., Hirano, S., Tsumoto, Y.: Similarity-based behavior and process mining of medical practices. Future Gener. Comput. Syst. **33**, 21–31 (2014). https://doi.org/10.1016/j.future.2013.10.014

22. Valerio, D.O., Santos, E.A.P., Loures, E.F.R., Cestari, J.M.A.P.: Application of process mining in after-sales on an automotive industry (2018). https://doi.org/10.12783/dtetr/icpr2017/17635

23. van der Aalst, W.: Data Science in Action. In: van der Aalst, W. (ed.) Process Mining. Springer, Heidelberg (2016). https://doi.org/10.1007/978-3-662-49851-4_1

24. van der Aalst, W.: Service mining: using process mining to discover, check, and improve service behavior. IEEE Trans. Serv. Comput. **6**, 525–535 (2013). https://doi.org/10.1109/TSC.2012.25

25. van der Aalst, W.M.P.: Process Mining: Discovery, Conformance and Enhancement of Business Processes. Springer, Heidelberg (2011). https://doi.org/10.1007/978-3-642-19345-3

26. van der Aalst, W., et al. (eds.): Business Process Management Workshops, pp. 169–194. Springer, Heidelberg (2011). https://doi.org/10.1007/978-3-642-19345-3

27. van der Aalst, W.M.P., van Hee, K.M., van der Werf, J.M., Verdonk, M.: Auditing 2.0: using process mining to support tomorrow's auditor. Computer **43**, 90–93 (2010). https://doi.org/10.1109/MC.2010.61

28. van der Aalst, W.M.P., Schonenberg, M.H., Song, M.: Time prediction based on process mining. Inf. Syst. **36**, 450–475 (2011). https://doi.org/10.1016/j.is.2010.09.001

29. van der Aalst, W.M.P., Weijters, A.J.M.M.: Process mining: a research agenda. Comput. Ind. **53**, 231–244 (2004). https://doi.org/10.1016/j.compind.2003.10.001

30. Willaert, P., Van den Bergh, J., Willems, J., Deschoolmeester, D.: The process-oriented organisation: a holistic view developing a framework for business process orientation maturity. In: Alonso, G., Dadam, P., Rosemann, M. (eds.) BPM 2007. LNCS, vol. 4714, pp. 1–15. Springer, Heidelberg (2007). https://doi.org/10.1007/978-3-540-75183-0_1

31. Yang, W.-S., Hwang, S.-Y.: A process-mining framework for the detection of healthcare fraud and abuse. Expert Syst. Appl. **31**, 56–68 (2006). https://doi.org/10.1016/j.eswa.2005.09.003

32. European Alternative Fuels Observatory. https://www.eafo.eu/. Accessed 20 May 2020

Author Index

Printed in the United States
by Baker & Taylor Publisher Services